HOME REPAIR AND IMPROVEMENT

PLUMBING

TIME®
LIFE
BOOKS

Other Publications
THE TIME-LIFE COMPLETE GARDENER
JOURNEY THROUGH THE MIND AND BODY
WEIGHT WATCHERS® SMART CHOICE RECIPE COLLECTION
TRUE CRIME
THE AMERICAN INDIANS
THE ART OF WOODWORKING
LOST CIVILIZATIONS
ECHOES OF GLORY
THE NEW FACE OF WAR
HOW THINGS WORK
WINGS OF WAR
CREATIVE EVERYDAY COOKING
COLLECTOR'S LIBRARY OF THE UNKNOWN
CLASSICS OF WORLD WAR II
TIME-LIFE LIBRARY OF CURIOUS AND UNUSUAL FACTS
AMERICAN COUNTRY
VOYAGE THROUGH THE UNIVERSE
THE THIRD REICH
MYSTERIES OF THE UNKNOWN
TIME FRAME
FIX IT YOURSELF
FITNESS, HEALTH & NUTRITION
SUCCESSFUL PARENTING
HEALTHY HOME COOKING
UNDERSTANDING COMPUTERS
LIBRARY OF NATIONS
THE ENCHANTED WORLD
THE KODAK LIBRARY OF CREATIVE PHOTOGRAPHY
GREAT MEALS IN MINUTES
THE CIVIL WAR
PLANET EARTH
COLLECTOR'S LIBRARY OF THE CIVIL WAR
THE EPIC OF FLIGHT
THE GOOD COOK
WORLD WAR II
THE OLD WEST

*For information on and a full description
of any of the Time-Life Books series listed above,
please call 1-800-621-7026 or write:*
Reader Information
Time-Life Customer Service
P.O. Box C-32068
Richmond, Virginia 23261-2068

HOME REPAIR AND IMPROVEMENT

PLUMBING

BY THE EDITORS OF TIME-LIFE BOOKS, ALEXANDRIA, VIRGINIA

The Consultants

Kenneth A. Long, a licensed master plumber, is co-owner and chief executive officer of Long's Corporation in northern Virginia and a past president of the Virginia Association of Plumbing, Heating, and Cooling Contractors. "Plumbing," says Long, " is the art of running pipe."

Jeff Palumbo is a registered journeyman carpenter who has a home-building and remodeling business in northern Virginia. His interest in carpentry was sparked by his grandfather, a master carpenter with more than 50 years' experience. Palumbo teaches in the Fairfax County Adult Education Program.

Mark M. Steele is a professional home inspector in the Washington, D.C., area. He has developed and conducted training programs in home-ownership skills for first-time homeowners. He appears frequently on television and radio as an expert in home repair and consumer topics.

CONTENTS

Home Plumbing Basics

Whether you are repairing a leaking pipe or adding a new washbasin, the task is simplified if you know how your home's plumbing is laid out and how to work with common pipe materials. These plumbing fundamentals, set forth in the pages that follow, will also help you to diagnose problems—and may enable you to fix the difficulty without an expensive visit from a professional.

Tool Kit for Plumbing

Elements of a Plumbing System

Conduits for Water and Wastes

Supply Pipes and Fittings
Drainpipes and Fittings
Measuring Pipes
Calculating Pipe Dimensions

Repairing and Replacing Pipe

Replacing Copper with Copper
Mating Plastic to Copper
Repairing Steel Pipe with Plastic
Fixing Plastic Pipe
Mending Cast-Iron Drainpipe

Extending Plumbing to a New Sink

Three Schemes for Supply and Drain Lines
Gaining Access to Old Pipes
Laying Out the New Plumbing
Making Connections
Running Pipe to the Fixture

A run of copper pipe repaired with plastic →

Tool Kit for Plumbing

Many of the tools necessary for plumbing repairs and improvements are multipurpose instruments, such as screwdrivers, pliers, hammers, and adjustable wrenches. With the addition of the few specialized tools shown here, you can be ready not only to meet most plumbing emergencies but also to install and replace pipes and fixtures.

Included in this tool kit are implements for loosening and tightening plumbing hardware, cutting and soldering pipe, and clearing clogged sink and toilet drains. There is no satisfactory substitute for any of these tools, which are designed for the hardware unique to plumbing or for work in awkward spaces, such as under the sink.

Seat wrench.
The repair of faucet leaks caused by worn valve seats *(pages 62, 77)* requires a seat wrench. With a square tip on one tapered end and an octagonal tip on the other, the wrench fits the two most common types of seats in a wide range of sizes.

Spud wrench.
The wide-spreading, toothless jaws on this wrench firmly grasp large nuts found on toilets and sinks. The jaws, which lock in place once opened to the desired width, are shaped to fit into tight spaces.

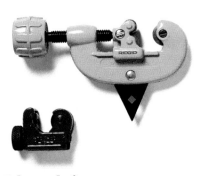

Tube and pipe cutters.
Depending on the type of cutting wheel installed, these devices cut either plastic or copper pipes. The built-in triangular reamer on the larger cutter scrapes away burrs around the cut edge, leaving it smooth. A minicutter is handy when working in tight spaces.

Pipe wrench.
The serrated teeth and spring-loaded upper jaw of this durable tool tightly grip pipes while you hold or turn them. The spring allows you to release the wrench's grip and reposition the tool without readjusting the jaws.

Basin wrench.
This self-adjusting tool's long handle is used primarily to reach otherwise inaccessible nuts that fasten faucets to washbasins and kitchen sinks.

Plunger.

A fold-out plunger has a flexible extension called a funnel. Extended as shown here, the funnel helps to unclog toilet drains. Folding it inside the cup converts the plunger for clearing tub, sink, and shower drains.

Faucet-handle puller.

To remove stubborn faucet handles, enlist the aid of this device. The jaws of the puller fit under the handle and pull it free as the threaded center shaft is tightened against the faucet stem.

Propane torch and flameproof pad.

To solder copper pipe or tubing—or to disassemble soldered joints—use a propane torch. With a flame spreader attached, the torch also thaws frozen pipes *(page 37)*. For any of these applications, protect nearby house framing with a flameproof pad.

Augers.

The trap-and-drain auger, shown here coiled with its handle, is used for clearing sink and tub drains; the handle slides along the snake as it progresses into the drain. For toilets, use a closet auger, which is shaped to direct the flexible shaft into the toilet trap and has a handle at the top.

Elements of a Plumbing System

Although fixtures and pipe materials vary, all plumbing systems share two basic components: a supply system to deliver water that is safe to drink and a drain-waste-vent (DWV) system to remove wastewater quickly and reliably.

System Basics: Water enters the house through a single pipe. This conduit passes through a water meter and then branches into hot- and cold-water supply lines, both of which carry water that is under pressure.

The DWV system includes drainpipes, which work by gravity, and vent pipes, which do not carry water; instead they allow gases to escape through the roof. Vents also equalize air pressure in the drains in order to prevent partial vacuums that could retard drainage. Underneath sinks, showers, and bathtubs—but inside toilets—water-filled bends that are known as traps prevent gases in the drains from entering the house.

Local Codes: Plumbing is controlled by local regulations that have the force of law. Observing these codes, besides being necessary, helps ensure the success of plumbing projects. For example, a roof vent is usually a single pipe, 3 to 4 inches in diameter. In the colder parts of Canada and the United States, snow and ice could block such a vent; codes there specify wider vent pipes.

Checking Water Use: Water meters show how many cubic feet of water are being consumed. Knowing how to read a water meter allows you to check how much water goes to a specific purpose such as lawn watering. Simply read the meter before and after the task, then take the difference. Convert readings from cubic feet to gallons with the calculator below.

Though some meters show a single easy-to-read figure as an odometer does, others have multiple harder-to-interpret dials *(bottom)*.

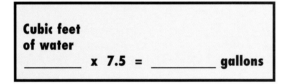

```
Cubic feet
of water
_____ x 7.5 = _____ gallons
```

Converting to gallons.
To translate water-meter readings in cubic feet to gallons, multiply the difference between two readings by 7.5.

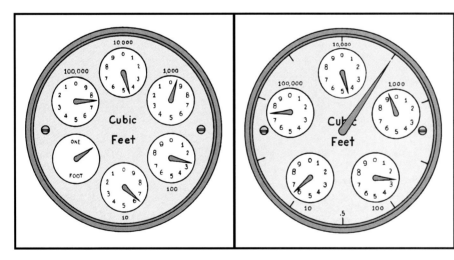

Reading a water meter.
On water meters with multiple dials, each is labeled with the number of cubic feet required to rotate the pointer a full turn. Thus each mark on the dial labeled 100,000 corresponds to 10,000 cubic feet of water. To take a reading, note the smaller of the two digits nearest the pointer, beginning with the 100,000 dial and ending with the dial labeled 10. The five digits from the dials provide the current reading (here, 74,926). Note that on a six-dial meter *(far left)*, the smaller digit lies clockwise from the hand on some dials, counterclockwise on others.

The dial labeled 1 on a six-dial meter—and the pointer that sweeps the edge of a five-dial model—measure fractions of a cubic foot, a feature helpful in detecting leaks. To confirm a leak, turn off all fixtures, then check whether this indicator on the meter continues to advance, however slowly.

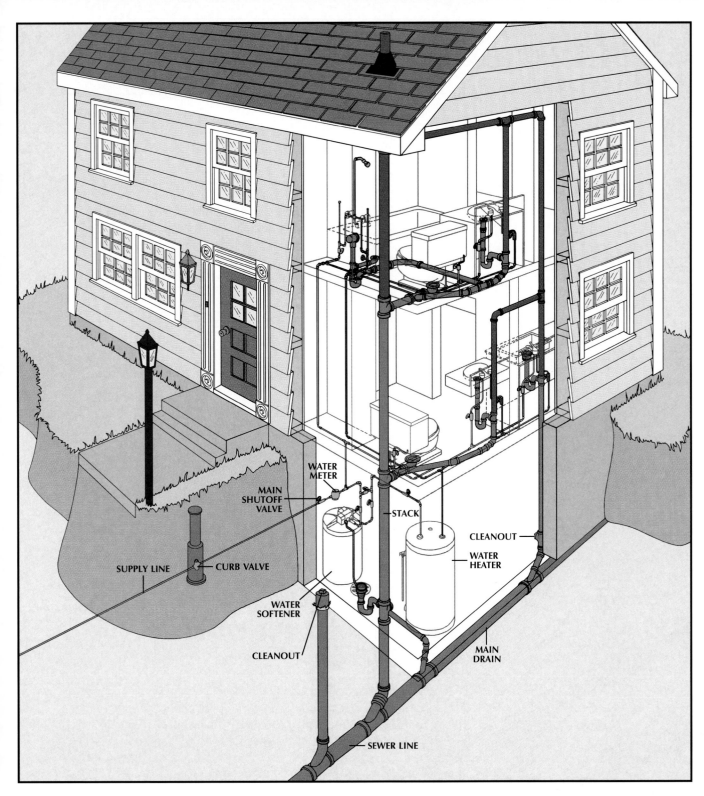

WATER METER

MAIN SHUTOFF VALVE

STACK

CLEANOUT

WATER HEATER

SUPPLY LINE

CURB VALVE

WATER SOFTENER

CLEANOUT

MAIN DRAIN

SEWER LINE

Anatomy of a plumbing system.

Typically, water reaches a house through a supply line controlled by two valves—an underground curb valve owned by the utility and the main shutoff valve in the basement. Past this valve is the water meter, beyond which the supply line divides. One branch supplies cold water to fixtures *(blue);*

the other supplies the water heater, often by way of a water softener *(pages 52-56).* Hot-water pipes *(red)* lead from the heater to the fixtures.

Drainpipes *(gray)* carry wastewater away from the plumbing fixtures to vertical stacks. These conduits lead to the house's main drain, which con-

nects to a sewer line. Cleanouts in both the stacks and the main drain provide access for unclogging the drain system.

Vent pipes *(purple)* channel gases from the drains through the upper portion of each stack and then outdoors through the roof.

11

Pipes and fittings, the prime constituents of a household plumbing system, not only can be assembled in any number of configurations but also exhibit considerable variety in their own right. For example, several different materials may be used for the supply pipes in residential systems, and still other materials are acceptable for the drain-waste-vent (DWV) portion of the network. (To determine what kind you have, simply examine the pipes in your basement, garage, or behind an access panel, and match them to the photographs on these pages.) As for fittings, they are designed for many roles—splicing straight lengths of pipe (called a run), allowing direction changes and branching, linking pipes of differing diameters, and so on. Some common sorts of fittings for the supply lines are shown opposite; DWV counterparts appear on page 14.

Keys to Buying Materials: Cost, durability, and ease of installation are among the important factors in choosing the materials for a repair job or an addition to your plumbing system. But before you make a purchase, check the code of your local jurisdiction. Some codes prohibit a particular material in one part of a plumbing system but not in another, and the local code may dictate the method used to join components. Among allowed materials, you can mix and match: Special transition fittings will create secure connections between dissimilar pipes; and if the pipes are made of different metals, an appropriate fitting will prevent an electrochemical reaction that could erode a joint.

All piping is sized by inside diameter. When replacing pipe, determine its inner diameter (pages 14-15) and buy new pipe of the same size.

SUPPLY PIPES AND FITTINGS

Rigid copper.
This metal, joined with durable solder, resists corrosion and has smooth surfaces for good water flow. A thin-walled version called Type M is the least expensive and will serve well for most repairs.

Chlorinated polyvinyl chloride (CPVC).
A rigid plastic formulated with chlorine so that it can withstand high temperatures, CPVC is a popular supply-pipe material for its low cost, resistance to corrosion, and ease of assembly: Fittings are secured by solvent cement. For residential use, most codes specify a so-called Schedule 40 pressure rating, stamped on the pipe.

Galvanized steel.
Used for supply as well as drain-waste-vent lines, this material is found only in older homes. Although it is the strongest of supply-pipe materials, galvanized steel is prone to corrosion over time. Runs of threaded pipe are joined by threaded fittings; the entire length between fittings must be removed and replaced to complete a repair.

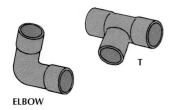

ELBOW   T

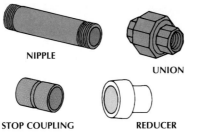

NIPPLE

UNION

STOP COUPLING   REDUCER

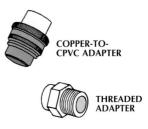

COPPER-TO-
CPVC ADAPTER

THREADED
ADAPTER

Bends and branches.

The direction of a supply pipe is changed by a fitting called an elbow, available in 45- and 90-degree turns. A T fitting joins a 90-degree branch run to a straight run of pipe.

Straight-line fittings.

Many sorts of fittings are used to join pipes in a straight run. A straight piece of pipe called a nipple extends a run or fitting a short distance—from 1 to 12 inches. A union holds two pipes within an assembly joined by a threaded nut, allowing disassembly without cutting. Couplings, unlike unions, are unthreaded and permanent. A stop coupling like the one shown here has interior shoulders for a secure fit in new installations; a slip coupling, which has no shoulders, slides over existing pipe and is used for repairs. A reducer, also known as a bushing, attaches pipes of different diameters by reducing the opening at one end.

Transition fittings.

As the name implies, these fittings allow a run of pipe to change from one material to another. Copper-to-CPVC adapters join the two most common kinds of supply pipe, both unthreaded. Threaded adapters join threaded pipe to unthreaded—and also are used to connect pipes to a variety of other plumbing components, such as a spigot or tub spout. Connecting steel pipe to copper requires another special-purpose fitting: a dielectric union, which prevents an electrolytic reaction between the two metals.

DRAINPIPES AND FITTINGS

Cast iron.

This is the strongest material available for DWV piping, and its heavy weight helps contain noise generated by active drains. Cast-iron pipe comes in two types, identified by the methods used to join them: hubless (left), which uses easily installed fittings; and hub and spigot, joined by a cumbersome procedure utilizing molten lead and oakum.

Copper.

Although mostly chosen for supply lines, copper pipe also comes in larger drain-waste-vent forms. Because copper DWV pipe is comparatively costly, however, it is seldom chosen for new installations.

ABS.

This plastic pipe, acronymically named for acrylonitrile butadiene styrene, is less expensive and more durable than PVC, but many codes prohibit its use because of low resistance to chemicals and a low ignition point. Where ABS is allowed in homes, a Schedule 40 pressure rating is recommended.

Polyvinyl chloride (PVC).

A rigid plastic pipe material like CPVC but less able to withstand heat, PVC is lightweight, easy to install, corrosion resistant, and inexpensive. In addition to its DWV uses, it can serve for cold-water supply lines. Codes usually specify a Schedule 40 rating for homeowners.

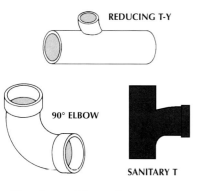

REDUCING T-Y

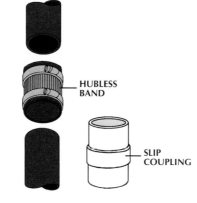

90° ELBOW

SANITARY T

HUBLESS
BAND

SLIP
COUPLING

THREADED
ADAPTER

FLEXIBLE
ADAPTER

Bends and branches.

A 90-degree elbow fitting (also known as a quarter-bend) makes a right-angle turn in DWV piping. Among the fittings that join two runs of drainpipe are a reducing T-Y, which connects a branch pipe to a larger diameter drainpipe, and a sanitary T, a fitting that joins a fixture drain to a vertical stack.

Straight-line fittings.

A hubless band—used to join hubless drainpipe—consists of a tightly fitting neoprene sleeve that is held in place over a joint by a stainless steel collar and clamps. A slip coupling, like its supply-system counterpart, slides over pipes to connect them in a repair.

Transition fittings.

A threaded adapter, as in the supply system, joins threaded pipe to unthreaded and also connects pipe to various special drain-system components, such as cleanout plugs and traps. A flexible adapter made of a rubberlike, specially treated PVC connects unthreaded drainpipes of any material; the fitting is slipped over the pipe ends and its built-in clamps tightened.

MEASURING PIPES

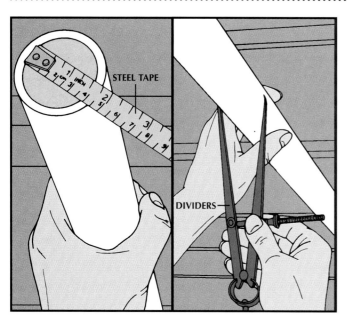

STEEL TAPE

DIVIDERS

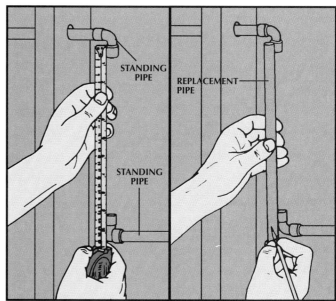

STANDING
PIPE

REPLACEMENT
PIPE

STANDING
PIPE

Finding the inside diameter.

Pipes and fittings are sized according to their inside diameter, called nominal size.

◆ To determine this figure for pipe that is already cut, simply hold a ruler or steel tape across an end of the pipe and measure from one inner wall to the other (above, left).

◆ If the pipe is part of an uncut run, you must proceed indirectly. First, fit dividers (above, right), calipers, or a C clamp against the pipe, then measure the space between the instrument's arms to get the outside diameter; repeat several times, and average the readings. Finally, convert the average outside diameter to inside diameter by referring to the chart opposite.

Measuring the replacement pipe.

◆ After cutting out or unthreading damaged pipe, buy the appropriate type of replacement pipe, making sure the piece is several inches longer than the gap. (See pages 17 to 25 to determine the correct pipe material and adapters for the plumbing being repaired.)

◆ Attach the new fittings on the ends of the standing pipes.

◆ Hold a steel tape to the farthest point the new pipe can extend into each of the two fittings (above, left). Alternatively, hold the replacement pipe up to the gap and mark the exact length—including the depth of the fittings—with a pencil (above, right).

CALCULATING PIPE DIMENSIONS

Reading the chart.

Always choose replacement pipe with the same inside diameter (ID) as the old pipe. To determine a pipe's inside diameter without cutting it, measure its outside diameter (OD) and use the conversion chart below. The chart also specifies the socket depth of fittings used with the various pipe sizes and types. For example, the socket depth of fittings for $\frac{3}{4}$-inch plastic pipe is $\frac{5}{8}$ inch. When you cut a length of replacement pipe, be sure to account for the fittings at both ends—twice $\frac{5}{8}$ inch, or a total of $1\frac{1}{4}$ inches.

COPPER

	Outside Diameter (OD)	Inside Diameter (ID)	Depth of Fitting Socket
Supply	$\frac{3}{8}$ in.	$\frac{1}{4}$ in.	$\frac{5}{16}$ in.
	$\frac{1}{2}$ in.	$\frac{3}{8}$ in.	$\frac{3}{8}$ in.
	$\frac{5}{8}$ in.	$\frac{1}{2}$ in.	$\frac{1}{2}$ in.
	$\frac{7}{8}$ in.	$\frac{3}{4}$ in.	$\frac{3}{4}$ in.
	$1\frac{1}{8}$ in.	1 in.	$\frac{15}{16}$ in.
Drains	$1\frac{3}{8}$ in.	$1\frac{1}{4}$ in.	$\frac{1}{2}$ in.
	$1\frac{5}{8}$ in.	$1\frac{1}{2}$ in.	$\frac{9}{16}$ in.
	$2\frac{1}{8}$ in.	2 in.	$\frac{5}{8}$ in.
	$3\frac{1}{8}$ in.	3 in.	$\frac{3}{4}$ in.
	$4\frac{1}{8}$ in.	4 in.	1 in.

GALVANIZED STEEL

	Outside Diameter (OD)	Inside Diameter (ID)	Depth of Fitting Socket
Supply	$\frac{3}{8}$ in.	$\frac{1}{8}$ in.	$\frac{1}{4}$ in.
	$\frac{1}{2}$ in.	$\frac{1}{4}$ in.	$\frac{3}{8}$ in.
	$\frac{5}{8}$ in.	$\frac{3}{8}$ in.	$\frac{3}{8}$ in.
	$\frac{3}{4}$ in.	$\frac{1}{2}$ in.	$\frac{1}{2}$ in.
	1 in.	$\frac{3}{4}$ in.	$\frac{9}{16}$ in.
	$1\frac{1}{4}$ in.	1 in.	$\frac{11}{16}$ in.
Drains	$1\frac{1}{2}$ in.	$1\frac{1}{4}$ in.	$\frac{11}{16}$ in.
	$1\frac{3}{4}$ in.	$1\frac{1}{2}$ in.	$\frac{11}{16}$ in.
	$2\frac{1}{4}$ in.	2 in.	$\frac{3}{4}$ in.

CAST IRON

	Outside Diameter (OD)	Inside Diameter (ID)	Depth of Fitting (If Not Hubless)
Drains	$2\frac{1}{4}$ in.	2 in.	$2\frac{1}{2}$ in.
	$3\frac{1}{4}$ in.	3 in.	$2\frac{3}{4}$ in.
	$4\frac{1}{4}$ in.	4 in.	3 in.
	$5\frac{1}{4}$ in.	5 in.	3 in.
	$6\frac{1}{4}$ in.	6 in.	3 in.

PLASTIC

	Outside Diameter (OD)	Inside Diameter (ID)	Depth of Fitting Socket
Supply	$\frac{7}{8}$ in.	$\frac{1}{2}$ in.	$\frac{1}{2}$ in.
	$1\frac{1}{8}$ in.	$\frac{3}{4}$ in.	$\frac{5}{8}$ in.
	$1\frac{3}{8}$ in.	1 in.	$\frac{3}{4}$ in.
Drains	$1\frac{5}{8}$ in.	$1\frac{1}{4}$ in.	$\frac{11}{16}$ in.
	$1\frac{7}{8}$ in.	$1\frac{1}{2}$ in.	$\frac{11}{16}$ in.
	$2\frac{3}{8}$ in.	2 in.	$\frac{3}{4}$ in.
	$3\frac{3}{8}$ in.	3 in.	$1\frac{1}{2}$ in.
	$4\frac{3}{8}$ in.	4 in.	$1\frac{3}{4}$ in.

A plumbing system's pipes, no matter what kind, are unlikely to remain problem-free forever. Sooner or later—perhaps because of corrosion, a leak at an aging joint, or the bursting of a frozen pipe—some mending will probably be necessary.

Measure for the replacement pipe as explained on page 14, and begin any supply-line repair by draining the system *(box, right)*. The methods for making repairs depend on the material of the pipes involved. You can use pipe and fittings that match your current piping, or introduce a different material—replacing a run of copper with CPVC, for example. Some of the most common ways of mending broken pipe are described opposite and on the following pages.

Copper Piping: Copper is connected to copper by heating the metal with a propane torch and drawing molten solder into a pipe-and-fitting joint—a process called "sweating." The solder for supply pipes must be lead-free; other solders are acceptable for copper drains. Because the flame can be a hazard to your home's structure and its wiring, cover the area be-hind the piping with a flameproof pad. Keep a fire extinguisher nearby, and turn off the torch before setting it down. For complex repairs, do as much of the assembly as possible at your workbench.

Connect copper to plastic with transition fittings; several types are available.

Plastic Piping: Assemble rigid plastic pipe only when the air temperature is above 40° F. Different types of plastic require their own primers and cements: Never use PVC primer on a CPVC repair, for instance. The basic methods used to join PVC and CPVC are the same, however *(page 23)*.

A third type of plastic, ABS, requires no primer. To repair ABS pipe with PVC replacement pipe, install a rubber adapter or consecutive male and female adapters in conjunction with primer on the PVC side and a light green transition cement on both sides.

⚠ **CAUTION** *Solvent cements are toxic and flammable. When you apply them, you must make sure the work area is well ventilated.*

SHUTTING DOWN THE SUPPLY SYSTEM

Before making repairs to the supply system, shut off the house water supply. First, close the main shutoff valve. Then, working from the top level down, open all hot- and cold-water faucets—including all tub, shower, and outdoor faucets—and flush all toilets. Open the drain faucets on the main supply line, the water heater, and any water treatment equipment you may have. To refill the system after the repair has been completed, close all the faucets, then open the main shutoff valve. Trapped air will cause faucets to sputter momentarily when you first turn them on again.

SAFETY TIPS

Goggles and gloves provide important protection when you are soldering copper pipe.

 TOOLS

Tube cutter	Flux brush	Fine-tooth
Hacksaw	Groove-joint	hacksaw or
Plumber's	pliers	minihacksaw
abrasive	Striker	Pipe wrenches
sandcloth	Propane torch	Ratchet pipe
Metal file	Flameproof pad	cutter
Round file	Small, sharp knife	Nut driver or
Wire fitting brush	Adjustable wrench	socket wrench

 **MATERIALS**

Replacement pipe and fittings	Applicator brushes
Clean cloth	Plumbing-sealant tape
Paste flux	Grounding clamps
Solder	Grounding wire
Miter box	2 x 4 lumber
Adapters	Stack clamps
PVC or CPVC primer	Chalk
PVC, CPVC, or ABS cement	Newspaper or paper towels

REPLACING COPPER WITH COPPER

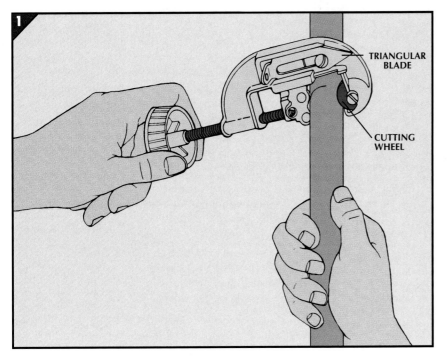

TRIANGULAR BLADE

CUTTING WHEEL

1. Using a cutter.

Although a hacksaw may be needed for a hard-to-reach section of broken copper pipe, use a tube cutter if possible.

◆ Slide the cutter onto the pipe and turn the knob until the tube cutter's cutting wheel bites into the copper. Do not tighten the knob all the way.

◆ Turn the cutter once around, retighten the knob, and continue turning and tightening. Once the piping is severed, loosen the knob, slide the cutter down the pipe, and cut through the other side of the broken section.

◆ With the cutter's triangular blade, ream out the burr inside the standing pipe. Remove the ridge on the outside with a file. (For a hacksaw cut, remove the inner burr with a round file.)

◆ Use the tube cutter to cut and ream replacement pipe.

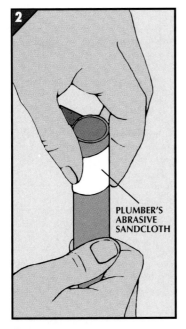

PLUMBER'S ABRASIVE SANDCLOTH

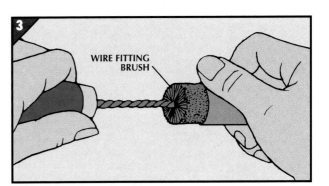

WIRE FITTING BRUSH

3. Cleaning the fittings.

Scour the inner surfaces of the sockets of each fitting with a wire fitting brush. Once the surfaces of the fittings and pipes have been cleaned, do not touch them: Even a fingerprint will weaken the joint.

2. Preparing the cut ends.

With a piece of plumber's abrasive sandcloth—not a file or steel wool—clean the cut pipe ends to a distance slightly greater than the depth of fittings that you will use to connect them. Rub until the surface is bright.

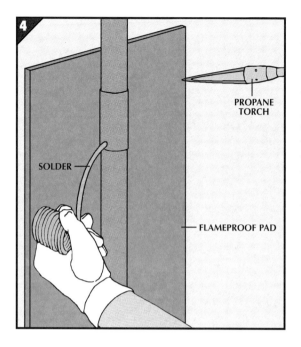

4. Assembling and heating a joint.

◆ Brush a light coat of flux over the cleaned surfaces, place the fitting between the standing pipe and the replacement section, and twist it a quarter-turn.

◆ For a slip coupling on a vertical pipe, shown here, gently crimp the coupling with groove-joint pliers just enough to hold it in place.

◆ Place a flameproof pad in back of the joint.

◆ Light the torch with a striker. Holding the tip of the flame perpendicular to the metal and about a half-inch away, play it over the fitting and nearby pipe.

◆ Touch a piece of solder to the fitting *(left)* until it melts on contact. Do not heat further or the flux will burn off and the solder will not flow properly.

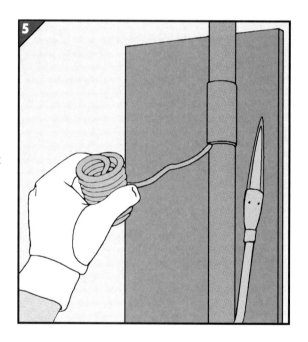

5. Soldering the joint.

◆ Touch the solder tip to the pipe where it enters one end of the fitting. Keep the solder at that point while the capillary action of the flux draws molten solder into the fitting to seal the connection.

◆ Remove the solder from the joint when a bead of metal completely seals the rim.

◆ Wipe away excess with a clean cloth, leaving a shiny surface.

◆ Apply solder to the other end of the fitting in the same way, then sweat the other joint.

Dos and Don'ts of Sweated Joints

Good sweated joints *(below, left)* are achieved by careful handling of flux and solder. Spread the coat of flux thinly and evenly. Excessive residue can cause corrosion; too little flux will create gaps in the bond between solder and copper. Do not overheat the fitting or direct the flame into the socket: If the flux burns—indicated by a brownish black coloring—the bond will be imperfect. Never direct the flame at the solder, and be sure to remove the solder as soon as capillary action sucks it around the full circumference of the joint. If solder drips, it has been overheated or overapplied, and the capillary action will fail. Thick, irregular globs of solder at the edges of sweated joints are a sign of a bad job *(below, right).*

MATING PLASTIC TO COPPER

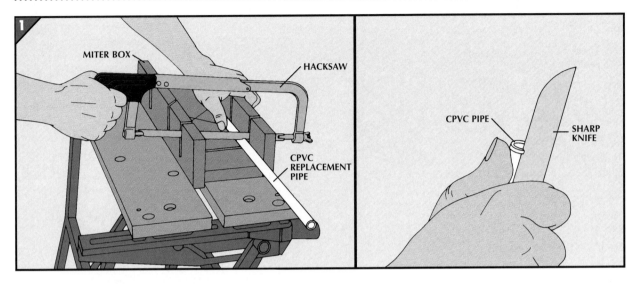

1. Cutting the pipes.

◆ Cut out the broken section of copper pipe with a tube cutter or hacksaw and ream the ends of the standing pipes (page 17).

◆ Cut the CPVC replacement pipe with a tube cutter or hacksaw. If you use a hacksaw, place the pipe in a miter box and brace it with your thumb as you make the cut (above, left).

◆ Ream the ends of the replacement pipe.

◆ With a small, sharp knife (above, right), trim the ends' inside edge to aid water flow and the outside edge to improve the welding action of the solvent.

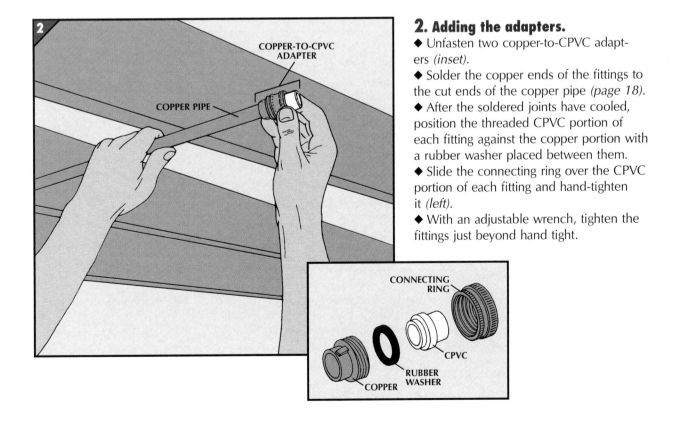

2. Adding the adapters.

◆ Unfasten two copper-to-CPVC adapters (inset).

◆ Solder the copper ends of the fittings to the cut ends of the copper pipe (page 18).

◆ After the soldered joints have cooled, position the threaded CPVC portion of each fitting against the copper portion with a rubber washer placed between them.

◆ Slide the connecting ring over the CPVC portion of each fitting and hand-tighten it (left).

◆ With an adjustable wrench, tighten the fittings just beyond hand tight.

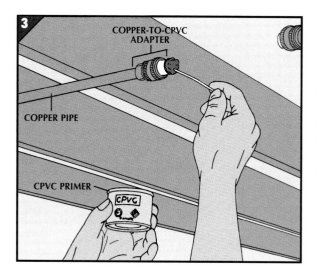

3. Priming and cementing the joints.

Work as quickly as possible with solvent cement. It sets in less than 30 seconds.

◆ With an applicator or clean cloth, apply a coat of primer to the inside of the sockets of the adapters *(left)* and to the outside pipe surfaces that will be fitted into the adapters.

◆ With a second applicator, spread a coat of CPVC cement over the primed surfaces at the ends of the CPVC pipe.

◆ Spread a light coat of cement inside the adapter sockets.

4. Fitting the replacement pipe.

◆ Working rapidly, push one end of the CPVC pipe into an adapter.

◆ Pull the free ends toward you until enough space opens for the CPVC pipe to slip into the second adapter *(right)*.

◆ Give the CPVC pipe a quarter-turn to evenly distribute the cement inside the sockets.

◆ Hold it firmly for about 10 seconds.

◆ Wipe away any excess cement with a clean, dry cloth. Do not run water in the pipe until the cement has cured (about 2 hours at temperatures above 60°F).

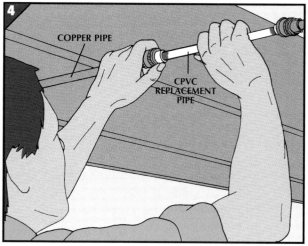

REPAIRING STEEL PIPE WITH PLASTIC

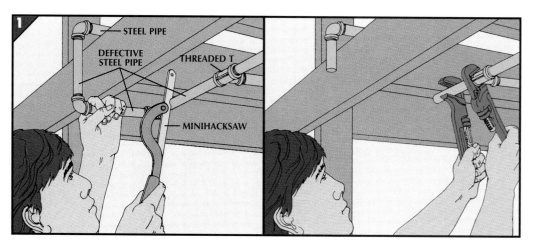

1. Removing the steel pipe.

Once in place, a threaded pipe cannot be unscrewed as one piece. In the situation here, CPVC replaces three runs of damaged pipe.

◆ Near the outer ends of the damaged section, cut the pipe with a fine-tooth hacksaw or minihacksaw *(above, left)*. Remove the intervening piping.

◆ Unthread the remaining stubs of pipe from their fittings. Hold a fitting stationary with one wrench and turn the pipe with another wrench. The jaws must face the direction in which the force is applied *(above, right)*.

If a union is near a damaged section, cutting is unnecessary. Hold the pipe steady with one wrench, unscrew the union with a second, then unscrew the other end of pipe from its fitting.

20

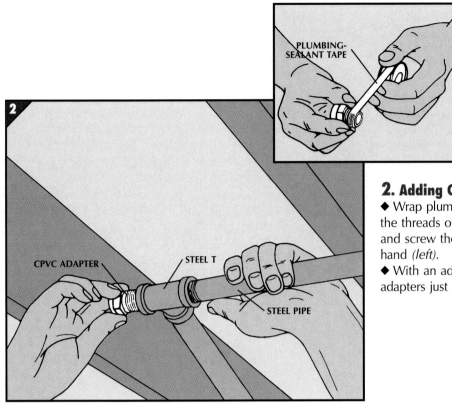

2. Adding CPVC adapters.

◆ Wrap plumbing-sealant tape around the threads of two CPVC adapters *(inset)* and screw them into the steel fittings by hand *(left)*.

◆ With an adjustable wrench, tighten the adapters just beyond hand tight.

3. Measuring and test-fitting replacement pipe.

◆ Push CPVC pipe into an adapter socket as far as it will go. Mark the desired length, allowing for the socket depth of the fitting that will go at the other end.

◆ Cut that section of pipe to length with a hacksaw in a miter box *(page 19)*.

◆ Push the fitting—in this case, an elbow—on the other end. Measure and cut the next length of CPVC pipe, push it into the next fitting, and continue dry-fitting the replacement piping in this way.

◆ At each connection, draw a line across the fitting and adjacent pipe *(right)* as a guide for reassembly and cementing.

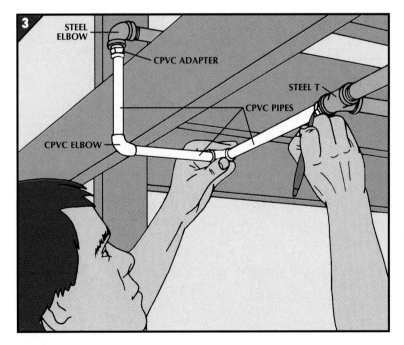

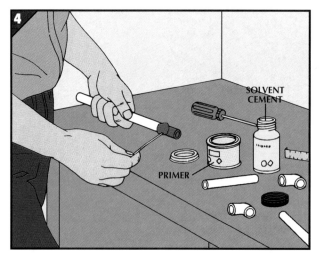

4. Cementing the CPVC pipe.

◆ Disassemble the dry-fitted pipe sections.

◆ At a well-ventilated workbench, ream and trim all pipe ends *(page 19)*.

◆ Apply primer and solvent cement *(page 20)* to a pipe and fitting that will form an outer section of the assembly *(left)*.

◆ Push the pipe into the fitting, give it a quarter-turn, and align the marks. Hold the pieces together for about 10 seconds.

◆ Continue cementing pipe and fittings together until all but the last pipe is in place. Leave this pipe detached.

5. Beginning the installation.

◆ Check to make sure that the CPVC adapters are dry. If they are not, dry them with a clean cloth.

◆ Apply primer to the sockets of both CPVC adapters.

◆ Spread solvent cement in the socket of the adapter that will receive the pipe at the completed end of the CPVC assembly. Apply primer and then a coat of cement to the end of that pipe.

◆ Push the pipe into the adapter socket as far as it will go. Give the pipe a quarter-turn to spread the cement.

◆ Line up the marks on the pipe and adapter and hold the pieces together for about 10 seconds *(right)*.

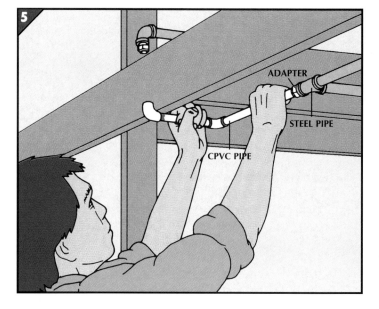

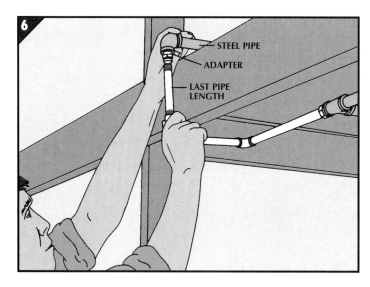

6. Adding the last pipe.

◆ Apply primer and cement to the second adapter, the last fitting on the CPVC assembly, and both ends of the unattached pipe.

◆ Push one end of the pipe into the assembly fitting. Gently maneuvering the assembly *(left)*, push the other end into the adapter socket as far as it will go.

◆ Give the pipe a quarter-turn to spread the cement, and hold the pieces together for about 10 seconds. Do not run water in the pipe until the cement has cured (about 2 hours at temperatures above 60°F).

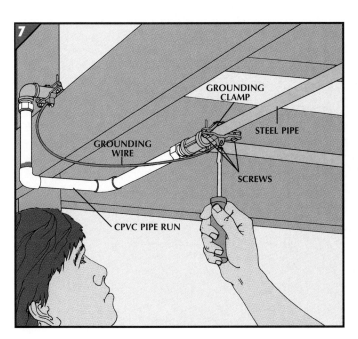

7. Installing a grounding jumper.

If the replaced section of steel pipe was part of your home's electrical grounding system, you must install a grounding jumper to maintain continuity.

◆ At one end of the cut steel pipes, fit both pieces of a grounding clamp *(photograph)* around the circumference of the pipe. Fasten the two pieces together with screws on either end of the clamp.

◆ Measure and cut a length of grounding wire to extend between the clamp and the other cut steel pipe.

◆ Insert the wire into the small opening on top of the clamp and secure it with the corresponding screw.

◆ Install a clamp on the other steel pipe and secure the end of the wire in its opening *(left)*.

FIXING PLASTIC PIPE

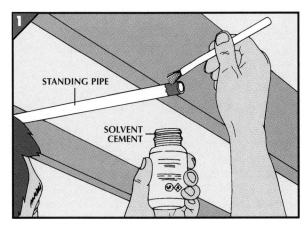

1. Preparing the joint.

◆ Cut out the damaged pipe with a hacksaw or tube cutter.

◆ Hold the replacement pipe against the gap and mark it, then cut it with a tube cutter or hacksaw and miter box.

◆ Ream and trim the pipe ends *(page 19)*.

◆ Prime the ends of the standing pipes and one socket on each of two couplings.

◆ Apply a liberal coat of solvent cement to the coupling sockets and on the ends of the standing pipes *(left)* to a distance matching the socket depth.

◆ Push the couplings onto the pipes, give them a quarter-turn to spread the cement, and hold the pieces together about 10 seconds.

2. Inserting the replacement pipe.

◆ Prime the exposed coupling sockets and ends of the replacement pipe, then apply cement.

◆ Working quickly, push one end of the replacement pipe into a coupling, then gently bend the pipes until the opposite end fits into the other coupling *(right)*.

◆ Give the pipe a quarter-turn to spread the cement, and hold the pieces together for about 10 seconds.

◆ Wipe off any excess cement around the pipe or fittings with a clean, dry cloth. Do not run water in the pipe until the cement has cured (about 2 hours at temperatures above 60°F).

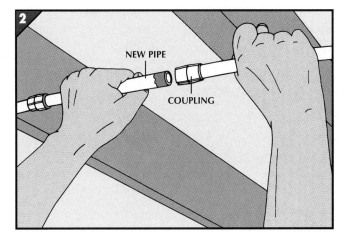

MENDING CAST-IRON DRAINPIPE

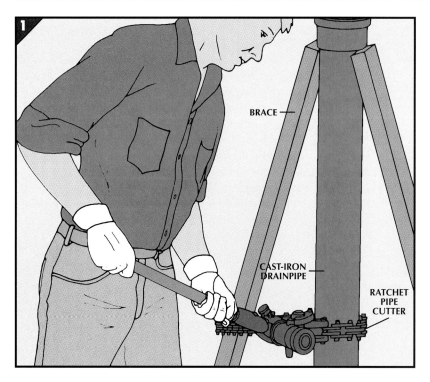

BRACE

CAST-IRON DRAINPIPE

RATCHET PIPE CUTTER

1. Removing the broken pipe.

The easiest way to cut cast-iron pipe is with a ratchet pipe cutter, available at rental stores. Before cutting vertical drainpipe, support it with stack clamps *(page 29)*, or brace 2-by-4s against a joint above the section to be removed *(left)*, hammering the braces into position for a secure fit. Make sure that no one runs water in the house during the repair.

◆ With chalk, mark off the area to be cut out.

◆ Wrap the chain around the pipe and hook it onto the body of the tool.

◆ Tighten the knob, turn the dial to CUT, and work the handle back and forth until the cutting disks bite through the pipe.

◆ If badly corroded pipe crumbles under a pipe cutter, rent an electric saber saw and metal-cutting blade.

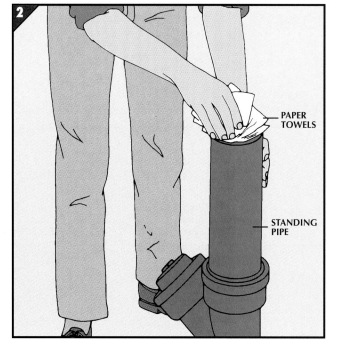

PAPER TOWELS

STANDING PIPE

2. Cutting the replacement pipe.

◆ Immediately after removing the damaged section, stuff newspaper or paper towels into the standing pipes *(left)* to block dangerous sewer gas.

◆ Measure the gap in the pipe and transfer that measurement, less $\frac{1}{4}$ inch, to a cast-iron, PVC, or ABS replacement pipe.

◆ Lay cast-iron pipe across two level 2-by-4s, spaced to support the pipe ends, and cut it to size with a ratchet cutter or saber saw.

◆ Cut PVC or ABS with a hacksaw and miter box.

◆ With a sharp knife, ream and trim the ends of the plastic pipe *(page 19)*.

3. Inserting the replacement section.

Hubless bands *(page 14)* join cast-iron drainpipe to replacement pipe.

◆ Slide a clamp onto each standing pipe and tighten the clamps to hold them temporarily in place.

◆ Slip the neoprene sleeves of the fittings onto each pipe until the pipe ends bottom out inside the sleeves *(near right)*.

◆ Fold the lip of each sleeve back over the pipe.

◆ Work the replacement pipe into the gap between the sleeves *(far right)* until it is properly seated.

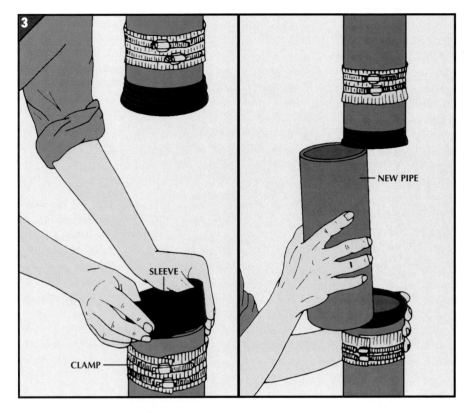

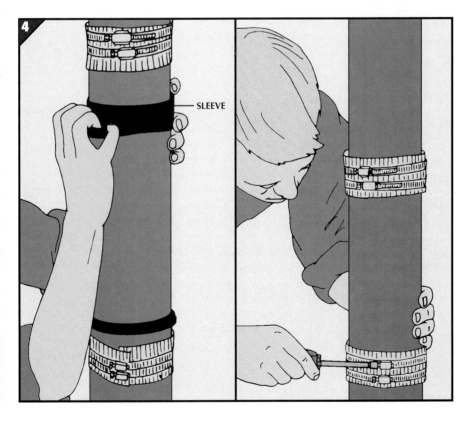

4. Completing the repair.

◆ Pull the folded lips of the sleeves over the replacement pipe *(far left)*.

◆ Loosen the clamps and slide them toward the replacement section until they are centered over the joints. Tighten them again with a nut driver *(near left)* or socket wrench.

◆ Run water through the drainpipe in order to test the repair; if a joint leaks, take it apart and reassemble the hubless band.

Extending Plumbing to a New Sink

With a little ingenuity, you can apply the techniques of cutting and connecting pipes *(pages 16-25)* to add a new fixture such as a washbasin or a wet bar to existing plumbing. In some situations, like the one illustrated below, you can complete the entire job outside the wall. Other projects may require cutting open and patching a small section. If the new piping runs exposed along a wall *(opposite, bottom)*, you can conceal it with one of the methods that are described on page 33.

For almost any extension, plastic pipes (CPVC for supply lines, PVC for drainpipes) are easiest to work with. Cut the old piping and add a branch fitting of the same material.

Join the new piping to it with the appropriate adapter *(pages 13-14)*.

Locating Old Pipes: Before starting work, you must find the stack inside the wall. Determine the stack's general location by observing where it exits the roof, then pinpoint it with an electronic stud finder. Supply lines, also called risers, may run beside the stack or take a more circuitous route. You can often find the risers with the stud finder—or turn on the water, one faucet at a time, and listen for the flow with your ear against the wall.

Keeping It Simple: Once you know where the plumbing is, plan

the job to avoid complications. For instance, always place the fixture within a few feet of the drain stack to avoid the need for a separate vent. Avoid adding a connection for a new sink drain to the stack below a washing machine or toilet connection; doing so would also require adding a new vent. Check your local code for other restrictions.

Make sure the drainpipe has room to slope properly and that the lower end will not be below the level of water standing in the trap. The placement of supply lines is less critical, but if the lines follow the same slope they will be easy to drain; keep the two lines about 6 inches apart.

TOOLS

Electronic stud
  finder
Drill with
  screwdriver and
  twist bits
Utility knife
Steel tape measure
Soil-pipe cutter
Tube cutter
Hacksaw
Propane torch
Flameproof pad
Dry-wall knife

MATERIALS

CPVC supply
  pipe
PVC drainpipe
Supply and
  drain fittings
Rubber sleeves
  and clamps
Pipe anchors
Adapters
Plumbing-
  sealant tape
Solder
Flux
PVC and CPVC
  primer and
  cement
Stack clamps
2 x 4s
Dry wall
$1\frac{1}{2}$-inch dry-
  wall screws
Fiberglass-mesh
  tape
Joint compound

SAFETY TIPS

When soldering adapters to existing copper pipe, wear gloves and eye protection.

THREE SCHEMES FOR SUPPLY AND DRAIN LINES

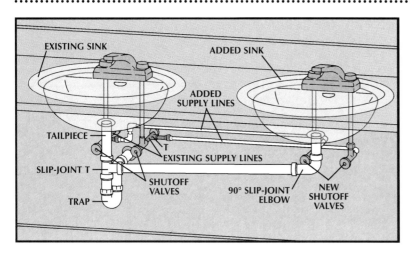

EXISTING SINK
ADDED SINK
ADDED SUPPLY LINES
TAILPIECE
EXISTING SUPPLY LINES
SLIP-JOINT T
SHUTOFF VALVES
TRAP
90° SLIP-JOINT ELBOW
NEW SHUTOFF VALVES

Side-by-side fixtures.

This arrangement requires no work on the stack—the new sink is tied into the trap of an existing one so that both empty into the stack together. The drainpipe between the two fixtures slopes downward from the new one $\frac{1}{4}$ inch per foot, and there can be no more than 30 inches between the drain holes of the two fixtures. Slip-joint fittings, including a slip-joint T above the trap of the old fixture, simplify the drain connections. The new supply lines run from T fittings added behind the old shutoff valves to the existing sink. At the added sink, a second pair of shutoff valves regulates water to that faucet.

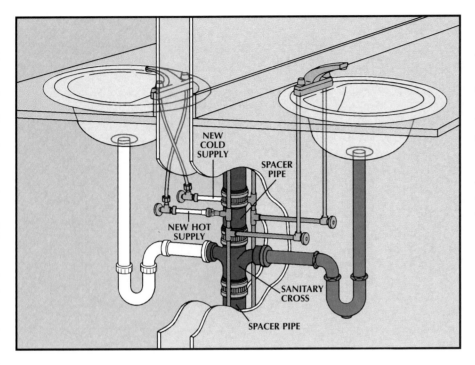

Back-to-back fixtures.

If an existing sink drains into a stack in the wall behind it, a new fixture can be installed back to back with the old one *(left)*. A two-inlet fitting called a sanitary cross, sandwiched between two spacer pipes, replaces the original drain connection, and the new supply lines run from T fittings added to the risers. Under the new sink, the supply tubes cross each other to bring hot water to the left side of the new fixture and cold water to the right.

NEW COLD SUPPLY

SPACER PIPE

NEW HOT SUPPLY

SANITARY CROSS

SPACER PIPE

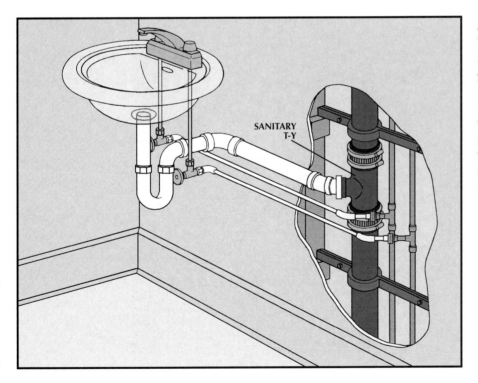

A fixture anywhere.

This method lets you install a fixture at any point close enough to a stack for a $\frac{1}{4}$-inch-per-foot drainpipe slope. A sanitary T-Y fitting or a sanitary T *(page 30)* makes the connection between the new drainpipe and the stack; supply lines are accommodated with T fittings on the risers. Both the drain and supply lines come outside the wall and run along it to the new fixture.

SANITARY T-Y

GAINING ACCESS TO OLD PIPES

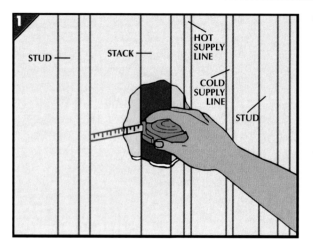

1. Uncovering the stack.
◆ Locate the stack with a stud finder, then verify its exact position by drilling an exploratory hole.
◆ With a utility knife, expand the hole so that the entire width of the stack is exposed.
◆ Insert a steel tape measure to find the distances from the hole to the studs on both sides *(left)*.

2. Opening the wall.
◆ Add $\frac{3}{4}$ inch to each measurement from Step 1 and use these distances to mark the middle of each stud on the wall; draw stud centerlines extending 12 inches above and below the hole.
◆ Between the tops and bottoms of the center-lines draw horizontal lines to mark the top and bottom of the opening.
◆ Cut the rectangle thus marked with a utili-ty knife. Set aside the cutout to use as a tem-plate later on.

If this opening does not uncover the supply lines, locate the lines and make a similar hole to ex-pose them.

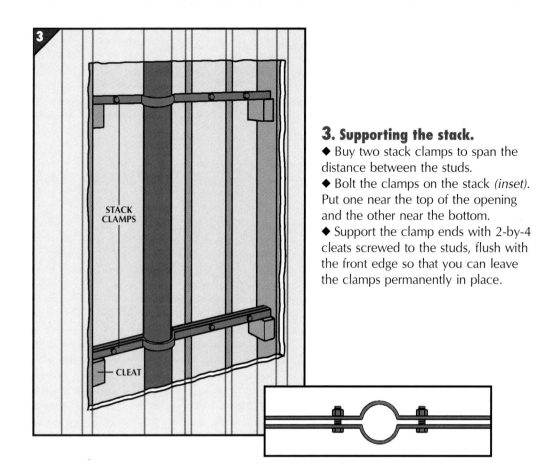

3. Supporting the stack.
◆ Buy two stack clamps to span the distance between the studs.
◆ Bolt the clamps on the stack *(inset)*. Put one near the top of the opening and the other near the bottom.
◆ Support the clamp ends with 2-by-4 cleats screwed to the studs, flush with the front edge so that you can leave the clamps permanently in place.

STACK CLAMPS

CLEAT

LAYING OUT THE NEW PLUMBING

COLD-WATER SUPPLY

HOT-WATER SUPPLY

TRAP EXIT

CENTERLINE

TRAP-EXIT HEIGHT

T-INLET HEIGHT

STACK

FIXTURE DRAIN HOLE

Roughing in.
◆ First, draw the centerline of the new fixture on the wall as a guide.
◆ Indicate the positions of the trap exit and the supply connections on the wall. On the floor, mark the location of the fixture drain hole.

◆ Transfer the height of the center of the trap exit to the stack. To indicate the center of the T inlet, lower the mark $\frac{1}{4}$ inch for each foot of the horizontal distance between the centerline and the stack *(above)*.
◆ Draw a drainpipe guideline between the trap-exit mark on the centerline and the T mark on the stack.

For a back-to-back installation *(page 27, top)*, reverse the sequence: Work from the existing stack connection to determine the height of the trap exit.

29

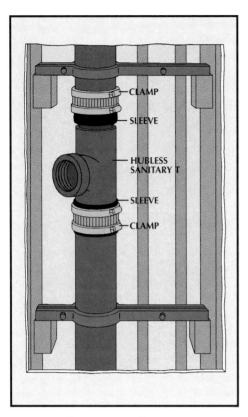

CLAMP

SLEEVE

HUBLESS SANITARY T

SLEEVE

CLAMP

A connection to cast iron.

◆ To add a single drain connection, align a hubless sanitary T with a threaded inlet at the inlet mark made earlier. Mark the stack at the top and bottom of the T.

◆ Cut the stack $\frac{1}{4}$ inch above and below these points *(page 25)*, then fit rubber sleeves and clamps onto the cut ends. Insert the T at a 45-degree angle to the wall and secure it with the sleeves and clamps *(above)*.

◆ For a back-to-back connection, disconnect the existing drainpipe; cut the stack about 3 inches above and below the T.

◆ Clamp a hubless sanitary cross between two pipe spacers *(page 27)*, positioning the cross at the correct height for connecting to the existing drain.

◆ Secure the assembly with the insets at right angles to the wall, then reconnect the drain.

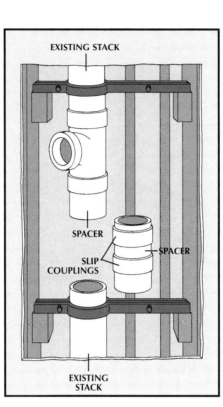

EXISTING STACK

SPACER

SPACER

SLIP COUPLINGS

EXISTING STACK

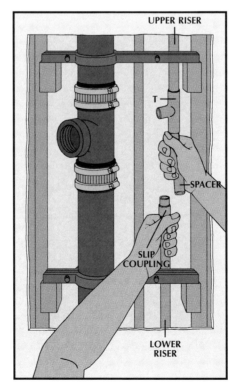

UPPER RISER

T

SPACER

SLIP COUPLING

LOWER RISER

A connection to copper or plastic.

◆ For a single drain connection, set a sanitary T against the stack at the height determined earlier. Cut the stack $\frac{1}{4}$ inch above and $\frac{1}{4}$ inch below the T.

◆ Solder or cement the T to the upper part of the stack, angling it as you would an iron one *(far left)*.

◆ Add a short spacer underneath the fitting.

◆ Cut a second spacer to fill the remaining gap; slide two slip couplings over it *(left)*. Set the spacer in place, slide the slip couplings over the joints, and solder or cement.

To create a back-to-back connection, detach the old drainpipe and cut the stack 3 inches above and below the T. Proceed as at left.

Adding a supply T.

◆ Drain the supply system as described on page 16.

◆ In copper or plastic pipe, cut an 8-inch opening for a T. Angle the T at 45 degrees to the wall and solder or cement it to the upper part of the riser. Slide a slip coupling onto the lower part.

◆ Cut a spacer for the gap and gently fit the spacer into the T *(left)*. Slide the slip coupling onto the lower end of the spacer and solder or cement it to seal the joint.

For back-to-back fixtures, install new Ts above or below the existing ones.

If the supply lines are steel, adapt the methods on pages 20 to 23 to insert a length of CPVC fitted with a T.

RUNNING PIPE TO THE FIXTURE

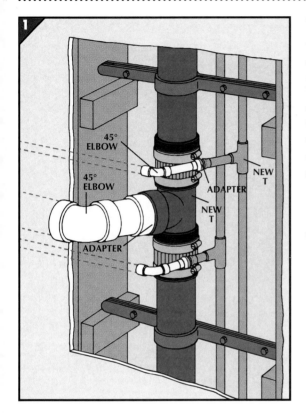

1. Getting outside the wall.

◆ Fit the openings of a sanitary T and supply-line T fittings with any necessary adapters *(pages 19 and 21)*. Then add CPVC supply stubs and PVC drain stubs just long enough to enter the room.

◆ Cement 45-degree elbows onto the ends of the stubs, rotating the elbows so that the new pipes *(dotted lines, left)* will point toward the new fixture.

Drainpipes and supply pipes in back-to-back installations extend directly into the room *(page 27)*.

2. Patching the wall.

◆ Trim a piece of dry wall to the same size as the section removed on page 28, Step 2. To accommodate pipe stubs and elbows, cut holes no more than an inch larger than the pipes.

◆ Fit the patch over the pipes, and use dry-wall screws to attach it to the studs.

◆ Cover the joints with self-adhesive, fiberglass-mesh tape *(left)*. Also apply tape over the gaps around each pipe.

◆ With a dry-wall knife, spread joint compound on the tape. After it dries, scrape away ridges with the knife. Apply a second coat, smooth the ridges again, and finish with a moist cloth.

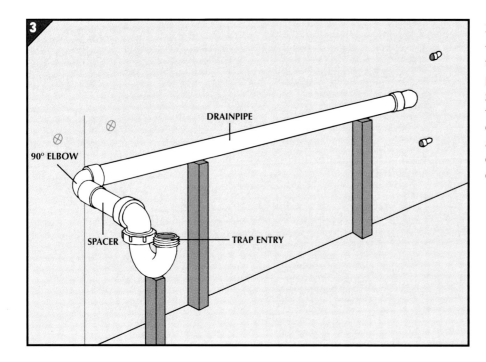

3. Running the drainpipe.

◆ Loosely assemble PVC drainpiping to reach the fixture location. Cut wood props to support the drain along the guideline marked on the wall *(left)*.

◆ Attach a 90-degree elbow at the end, and complete the assembly with a spacer and a trap to position the trap entry directly above the drain mark on the floor.

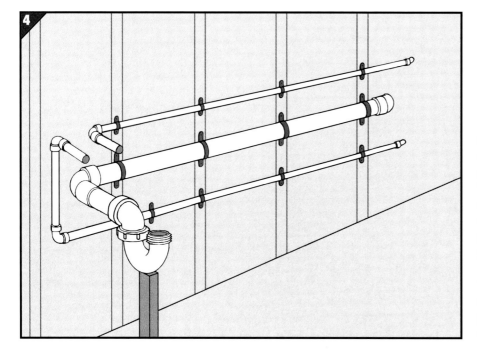

4. Anchoring the pipes.

◆ When you are sure everything fits in the drain assembly, cement the joints. Then anchor the pipe to every stud with metal straps.

◆ Run and anchor CPVC supply pipes as you did the drain. Fit the ends with 90-degree elbows and short spacers.

The fixture can now be installed and connected—to the trap with a tail-piece, to the supply lines with shutoff valves and supply tubes *(page 81)*.

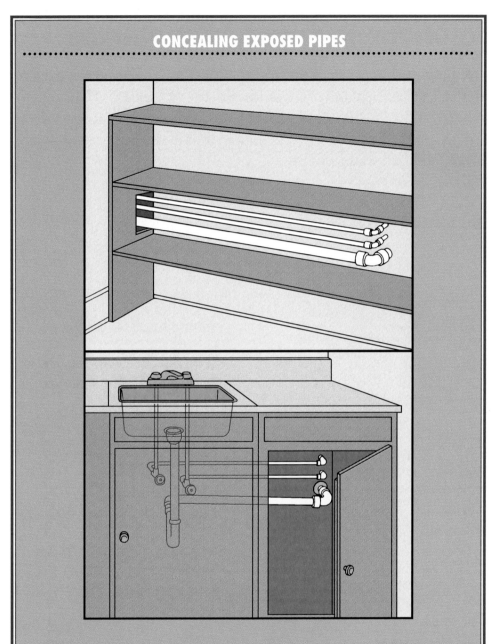

Runs of pipe are often left in view in utility or laundry rooms, and some people paint piping in other locations with bright colors and consider it a decorative asset. In many cases, however, you may prefer to hide the new fixture's piping. This is easy to do with shelves, boxes, cabinets, or closets, adapted to your particular situation.

To conceal a horizontal stretch of drainpipe and supply lines like the one depicted on the preceding pages, you can frame a box out of 1-by-2 lumber and complete the sides with plywood. The box can then be painted or finished with wallpaper. Bookshelves do the job even more simply *(top)*, since the books on the shelves will hide the pipes. Another common solution is to set the added fixture in a cabinet— known as a vanity in the case of a bathroom *(bottom)*.

Solutions to Common Problems

The following pages include suggestions for how to avoid frozen pipes, clogged drains, and other plumbing emergencies—as well as how to cope if the worst does happen. Also consult this chapter to correct some equally serious, although less urgent, plumbing problems: pressure that is too high or too low, excessive water usage, and bad water quality.

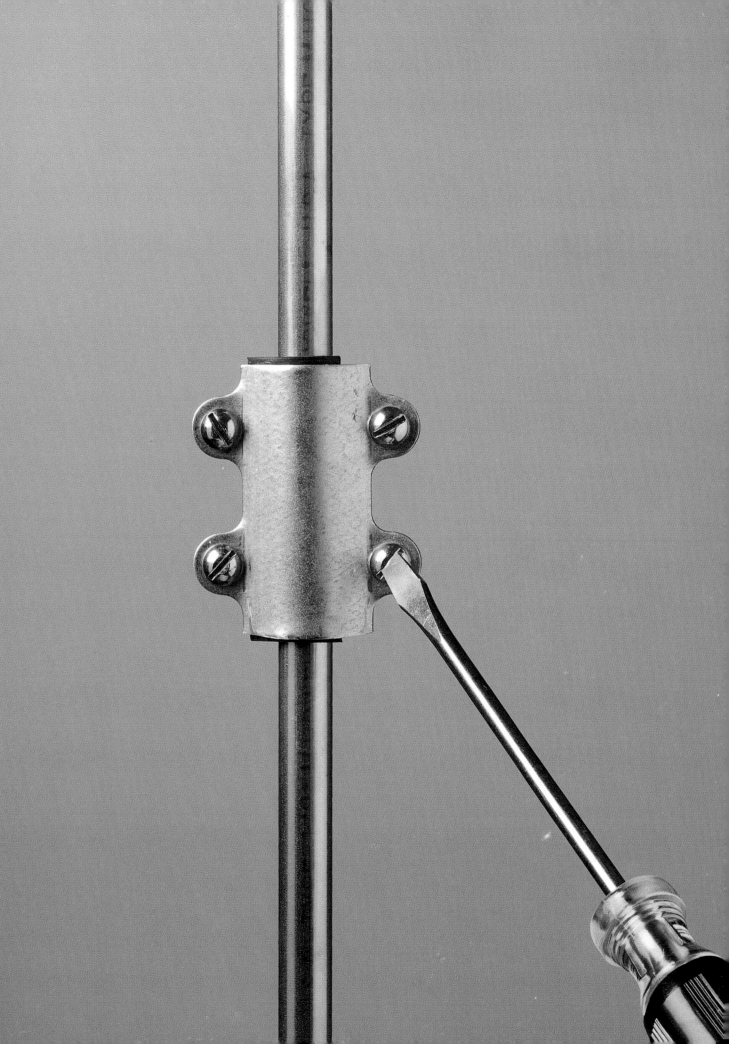

First Aid for Frozen Pipes

A house that is properly constructed and heated is safe from plumbing freeze-ups even in the midst of a severe cold snap—unless the heating system breaks down. If that should happen, the best way to keep pipes from freezing and bursting is to drain the entire plumbing system *(page 16)*. Also drain the plumbing in a house that will be left empty for the winter *(page 38)*.

Although a house may be well built, if its pipes run through a basement, crawlspace, laundry room, or garage that is unprotected, they may be vulnerable to cold. To avoid resulting problems, consult the checklist at right.

Coping with Leaks: If a pipe freezes despite your precautions, the first symptom may be a faucet that refuses to yield water. But all too often, the freeze-up is announced by a flood from a break. Ruptures are especially likely near joints or bends in the plumbing. When a leak occurs, turn off the water supply and apply a temporary patch *(page 39)*.

Getting Ready to Thaw Pipes: As you prepare to warm a frozen section of pipe, close the main shutoff valve most of the way. The movement of water through the pipe aids thawing and helps protect against later refreezing. Keep the affected faucet open to let water vapor and melted ice run out. Since leaks may go undetected until the pipe thaws, guard against water damage by spreading plastic drop cloths, and have extra pots and pails ready. Then warm the pipe by one of the methods at right and on page 38.

Electrical heaters of one kind or another are generally safest for thawing both metal and plastic pipe. Since electricity and water together pose a shock hazard, plug the appliance into a GFCI-protected outlet, which cuts power to the appliance if it detects conditions that could lead to injury.

TOOLS

Electric heating
  tape
Propane torch with
  flame spreader
Hair dryer
Heat gun
Heating pad
Heat lamp
Work lamp

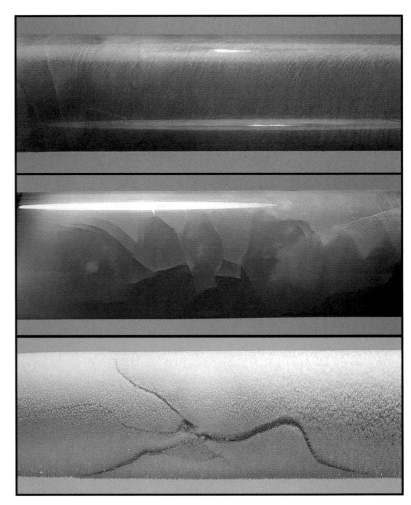

Three steps to a ruptured pipe.

Unlike most substances, water expands when it freezes—a fact that can easily burst a pipe. Three stages in the freezing and rupture of a pipe are shown in the transparent tubing at right. Frost forms first on the inner surfaces of the pipe *(top photograph)*, then ice crystals begin to take shape *(middle)*. With freezing complete, the pipe cracks *(bottom)*. By melting freeze-ups quickly, you may be able to avoid the final stage—and preserve your supply pipes from further harm.

How to Keep Pipes from Freezing

✔ Protect exposed pipes ahead of time with insulation made to retard freezing, or warm them with thermostatic heating tape *(right)*.

✔ When no commercial insulation is at hand and pipes must be protected immediately, wrap several layers of newspaper loosely around the pipes and tie the paper on with string.

✔ If you have no time to install insulation, open faucets so a trickle of water moves through the pipes.

✔ Keep a door ajar between a heated room and an unheated room with pipes so that the unprotected area will receive heat.

✔ If power is available, plug in an electric heater or heat lamp, or hang a 100-watt bulb near vulnerable pipes. Keep the heat source a safe distance from walls, floors, ceilings, and nearby combustibles.

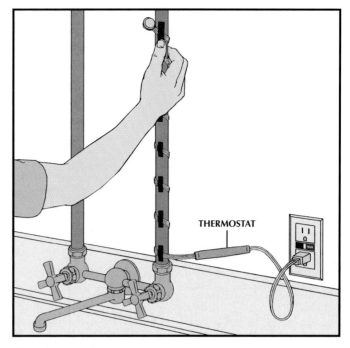

Electric heating tape.

To thaw a frozen pipe, wrap the tape in a spiral around the pipe, allowing about two turns per foot. Secure the spiral with PVC-type electrical tape *(above)*, which will stay in place during temperature changes.

Most electric heating cables come with built-in thermostats and can be left plugged in permanently: When the temperature drops toward freezing, the thermostat activates the cable and warms the pipe. Cover the pipe and heating cable with nonflammable fiberglass pipe insulation as a second layer of protection against freezing.

A propane torch.

Equipped with a flame-spreader attachment, available at most hardware stores, a propane torch can thaw metal pipes rapidly and effectively during a power outage, if used with care. Place flameproof sheeting between the pipe and nearby framing. Apply heat near an open faucet first, then work gradually along the pipe *(arrow)*.

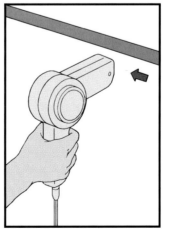

A hair dryer.

If you have power, use a hair dryer instead of a propane torch; the dryer will work more slowly, but you will avoid dealing with an open flame in close quarters. An electric heat gun can also be used to thaw pipes. As with a torch, make sure pipes never become too hot to touch.

⚠ **CAUTION** *Use a torch only on metal pipes, not plastic ones. Never let the pipe get too hot to touch; boiling water and steam inside a pipe can cause a dangerous explosion.*

A heating pad.
Wrapped and tied around a frozen pipe near an open faucet, an ordinary heating pad can be left in place to thaw ice slowly but effectively.

A heat lamp.
If a suspected ice blockage is behind a wall or above the ceiling, set an electric heat lamp nearby. Keep it at least 6 inches from the wall to avoid scorching paint or wallpaper. For greater flexibility in handling, you can screw the bulb into the socket of a portable work lamp *(above)*.

WINTERIZING AN EMPTY HOUSE

When you leave a house empty and unheated for the winter, take steps to weatherproof the plumbing. First turn off water to the house. Cut power to the water heater. For a hot-water heating system, turn off power to the boiler and drain it. Next, open the radiator valves, and remove an air vent from a radiator on the top floor.

Then empty the rest of the plumbing, including the water heater and any water-treatment devices, as described on page 16. For a well system, drain the storage tank and dry off the pump, unless it is submerged in the well.

Flush and bail out each toilet. Then pour at least a gallon of plumber's antifreeze—not the toxic automotive variety—into the tank and flush the toilet again. Doing so frostproofs both the trap and the flushing channels.

For other fixtures, pour antifreeze down the drain very slowly so that it displaces water in the trap rather than mixing with it.

Often the first sign of a leaking pipe is a spreading stain on a wall or ceiling or a puddle on the floor. Before trying to trace the leak, shut off the water supply to prevent further damage and to reduce pressure on the damaged section so you can repair the hole.

Minimizing Water Damage: Where leaking pipes are concealed above the ceiling and a water stain is visible, place a waterproof drop cloth on the floor and position a basin under the wet area. Poke a hole through the ceiling or remove a section of it to let any remaining water drain out—and stand out of the way! To deal with water leaking from a ceiling light fixture, shut off the electricity, then drain the fixture by removing its cover.

If you find a leak too late to avert a flood, construct a makeshift dam from rolled-up rugs to prevent water from spreading to other rooms. For a bad flood, you may need to rent a pump with a submersible motor. If the situation is desperate, call the local fire department.

Patching the Leak: Purchase a pipe-repair sleeve *(below)* to make a secure temporary patch. To make permanent repairs, replace the leaking section of pipe *(pages 16-25)*.

For a drainpipe, leaks are likeliest at the joints; sometimes a lead joint can be resealed as shown below at right. Otherwise, as with a supply pipe, replace the leaking pipes and joints.

TOOLS

Screwdriver   Hammer
Wrench   Chisel

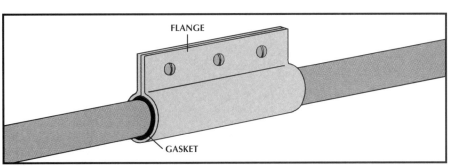

FLANGE

GASKET

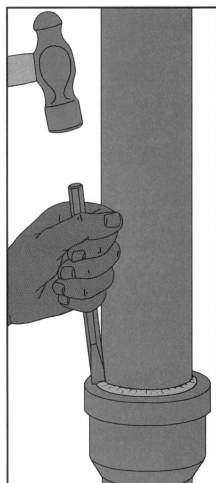

Installing a pipe-repair sleeve.
Measure the outside diameter of the leaking pipe *(page 12)*, then at a hardware or plumbing-supply store buy a temporary pipe-repair sleeve to fit. The pipe-repair sleeve seals a leak by means of a rubber gasket. Spread apart the flanges of the sleeve and then slip the sleeve around the damaged section, making sure you turn the flanges away from the hole. Finally, tighten the sleeve with a screwdriver or a wrench.

Fixing a lead-caulked drain joint.
If water seeps from a lead-caulked drain joint, tamp down the lead inside the hub of the pipe with a hammer and cold chisel *(right)*. Since the lead is soft enough to be reshaped over a weak spot, this procedure often reseals the joint.

When a sink empties slowly or not at all, the cause is usually debris blocking the drainpipe or trap just below. If other drains show the same symptoms, the problem is in the main drain or its branches *(below)*.

A Choice of Methods: To unclog a drain, try the simplest remedy first, then proceed stepwise with the progressively more demanding procedures shown on pages 41 and 42. A force-cup plunger offers the easiest way to loosen an obstruction and flush it away. A more powerful alternative is a stream of water delivered by a drain flusher *(page 42)*.

Should the obstruction resist these measures, try a trap-and-drain auger, or snake. This flexible steel coil is twisted into the pipes like a corkscrew to dislodge debris.

Some blockages may require that you open a sink trap or a cleanout in the main drain. Before doing so, turn off water to the entire house *(page 16)* and flush all toilets. Since wastewater may be trapped above a

cleanout, be ready for a dirty torrent as you remove the plug that seals the cleanout opening. If you have no success with the snake, call a plumber, who can bring an electric-powered auger that is capable of dealing with the most stubborn obstruction.

Chemical Solutions:
Drain cleaners may work effectively against a slow-flowing drain, and routine use every few weeks helps to prevent future clogs. But resist the temptation to pour these agents through standing water into a drain that is completely clogged. Extended contact with the lye or acid in drain cleaners can damage pipes and fixtures. Also, if you must later open a trap or a cleanout, drain cleaner in the pipes may splash on exposed skin or furnishings.

⚠️ **CAUTION** *Do not use a plunger, auger, or chemical cleaner in a sink equipped with a garbage disposer. Instead, disconnect the disposer's outlet pipe and clear obstructions as shown on page 42, Step 4.*

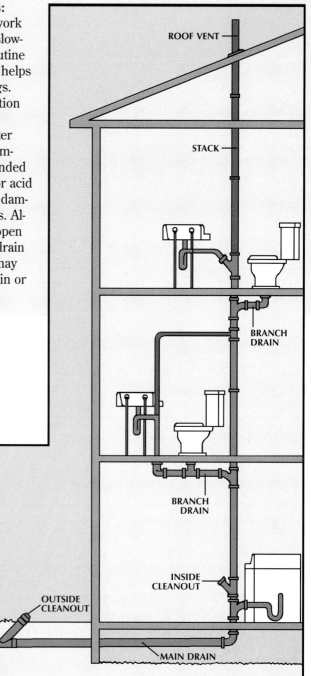

Finding the clog.
To clear a blocked pipe, pinpoint the obstruction by observation and deduction, then work from the drain—or cleanout, if available—immediately above.

For example, if the second-floor sink in the diagram at right is stopped but the toilet is clear, the clog is in the sink trap or the branch drain between the two fixtures. When all second-floor fixtures drain properly but all top-floor fixtures are stopped up, the clog is in the stack between floors. If all the drains back up, the problem is in the main drain. Most houses have only one cleanout plug for the main drain. It is near the foot of the stack in old houses; newer systems have an exterior cleanout within 5 feet of the foundation.

 **TOOLS**

Force-cup plunger
Trap-and-drain
 auger
Adjustable pipe
 wrench
Bucket
Bottle brush

 **MATERIALS**

Mop, rags, and
 sponges
Plastic bag
Petroleum jelly
Electrical tape
Plumbing-sealant
 tape
Penetrating oil

 **SAFETY TIPS**

Wear rubber gloves and safety goggles when using drain cleaners or working through a cleanout in a stack or main drain.

OPENING A SINK DRAIN

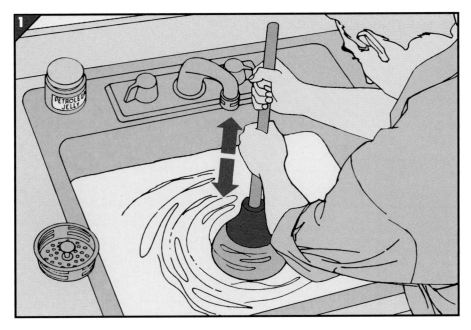

1. Starting with a plunger.

◆ Remove the sink strainer. In a double sink, bail out one side—the side with the garbage disposer, if there is one—and plug the drain with a rag in a plastic bag. In a washbasin, remove the pop-up drain plug; most lift out or can be turned and then lifted. Make sure there is enough water in the sink to cover the base of the plunger cup.
◆ Spread petroleum jelly on the cup's rim. Lower the plunger at an angle and compress the cup to push out air. Then seat it over the drain. Without breaking the seal, pump the plunger up and down 10 times, then quickly pull it away. If the drain stays clogged after several attempts, try an auger *(below)*.

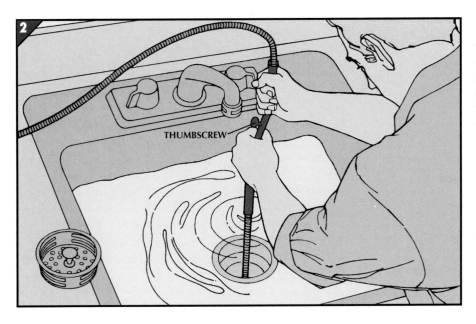

THUMBSCREW

2. Using an auger.

◆ Push the tip of the auger into the drain until you feel it meet the obstruction, then slide the handle within a few inches of the drain opening and tighten the thumbscrew.
◆ Crank the handle clockwise with both hands while advancing the auger into the drain. Continue cranking and pushing the auger, repositioning the handle as necessary, until the auger breaks through the obstruction or will not go any farther.
◆ Withdraw the auger by cranking it slowly and pulling gently.

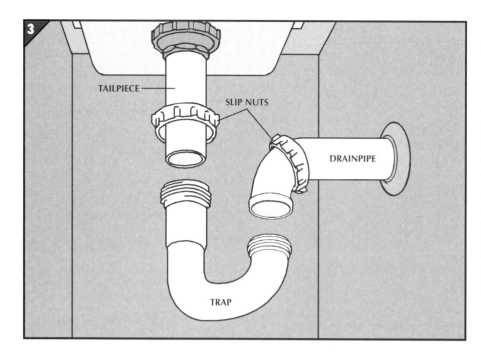

3. Removing the trap.

◆ Bail out the sink and place a bucket under the trap. Wrap the jaws of a pipe wrench with electrical tape, then loosen the slip nuts holding the trap to the tailpiece and drainpipe. Lower the trap slowly, allowing water to run into the bucket.

◆ If you find an obstruction in the trap, remove it, then clean the trap and the drainpipe with a bottle brush and detergent solution. Wrap the threads at both ends of the trap with plumbing-sealant tape, then replace the trap and tighten the slip nuts.

◆ If the trap was not blocked, clear the branch drain *(below)*.

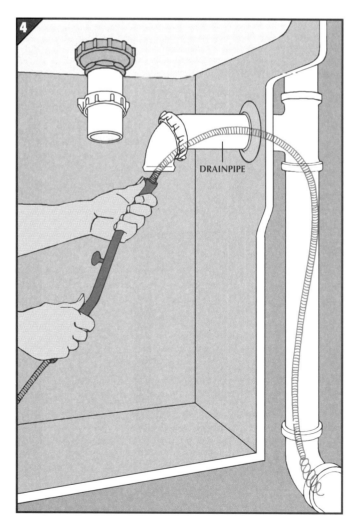

TRICKS OF THE TRADE

A Hydraulic Flusher

Concentrated water pressure can clear drains with a drain flusher *(right)*. Connect it to a faucet or hose and insert it into a drain. When water is turned on, the flusher expands and directs a jet of water down the drain. Because of potential damage to corroding metal pipes, do not use a flusher in older houses.

4. Cleaning beyond the trap.

With the trap removed, crank the auger into the drainpipe. The blockage may be in the vertical pipe behind the fixture or in a horizontal pipe—a branch drain—that connects with the main drain-vent stack serving the entire house. If the auger goes in freely through the branch drain until it hits the main stack, the blockage is probably in the main drainage system. In that case, identify the most likely location of the blockage *(page 40)*, open the drain just above it, and use the auger to clear the pipe.

CLEARING THE MAIN DRAIN

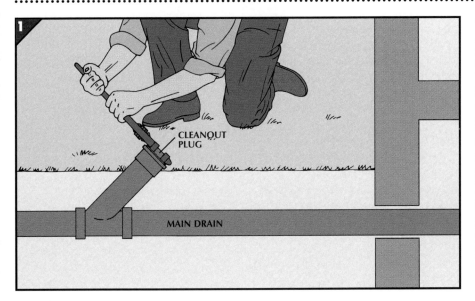

CLEANOUT PLUG

MAIN DRAIN

1. Opening the main cleanout.

Look for an outdoor cleanout at ground level within 5 feet of the foundation. The cleanout plug may be screwed into a Y fitting *(left)* or into the top of a vertical pipe that descends to a deeply buried main drain. Unscrew the cleanout plug with a pipe wrench. If the plug does not turn, apply penetrating oil around the perimeter, then try again. Depending on the extent of corrosion, the oil may take several days to free the plug. To gain more leverage, slip a section of pipe over the wrench handle to increase its effective length.

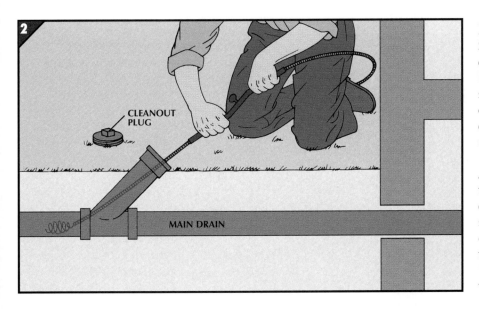

CLEANOUT PLUG

MAIN DRAIN

2. Working through the cleanout.

Standing water in the cleanout or drainpipe indicates that an obstruction lies between the cleanout and the sewer. Work an auger toward the sewer to clear the blockage *(left)*. If this does not work, recap the cleanout and notify a plumber.

An absence of standing water reveals that the obstruction is in the main drain under the house, or in the lower stack. Grease the threads of the cleanout plug and recap the pipe. Find the lowest drain opening or cleanout in the house and clear the blockage with an auger.

Increasing the Flow of Water

When the kitchen sink takes too long to fill, or the force of a shower is inadequate, the cause may be nothing more than a clogged faucet aerator *(page 72)* or shower head *(page 69)*. If water flow is weak at several fixtures, however, the problem lies in the supply system.

Low Pressure at the Source:
First make sure that the main shutoff valve is fully open *(page 11)*, then check the water pressure *(below)*. Most municipal systems provide water at a pressure of at least 50 pounds per square inch. For a lower reading, ask your neighbors whether their pressure is satisfactory. If so, there may be a leak in the pipe between the city main and your house. When the municipal system itself has low pressure, a pump and pressure tank hooked to the main supply line may be the best solution *(page 47)*.

In a home served by a well, the source of the problem may be a low setting on the pump's pressure switch. You can increase pressure by adjusting the switch *(page 47)*.

Blockages in the Plumbing:
Where water pressure to the house is adequate, compare the main shutoff valve with the diagrams on the next page. If it is a globe valve, a type that restricts water flow, replace it *(page 46)*.

With the flow-rate chart at right, you can determine what areas of the house may have pipes or valves that have become clogged over time with mineral deposits. Clean or replace clogged valves *(page 46)*. As a last resort, consider installing new pipes *(pages 16-25)*.

TOOLS

Pressure gauge	Adjustable wrench
Graduated bucket	Wire brush
Stopwatch	Propane torch
Pipe wrench	Flameproof pad
Small open-ended wrenches	Pliers (two pairs)

MATERIALS

Penetrating oil	Flux
Plumbing-sealant tape	CPVC primer and cement
Solder	

SAFETY TIPS

When soldering copper pipe joints or heating and loosening old ones, wear gloves and safety goggles.

DIAGNOSING THE PROBLEM

Measuring water pressure.
If you have a well, simply read the gauge at the pump *(left)*. Check municipal pressure as follows:
◆ Open the main shutoff valve fully and turn off all faucets and appliances.
◆ Screw a pressure gauge, which is available from hardware stores and lawn sprinkler suppliers, to the faucet of a basement laundry tub or an outdoor sillcock *(inset)*.
◆ Turn on the faucet or sillcock all the way and read the gauge. Take readings at different times of the day, since municipal pressure varies according to volume of use.

44

MINIMUM FLOW RATES	
Fixture	Gallons per minute
Laundry tub	5
Kitchen sink	4.5
Bathroom basin	2
Bathtub	5
Shower	5
Sillcock	5

Taking a flow-rate test.

This chart lists flow rates at various points in a residential plumbing system. To check the flow rate at a fixture, remove any aerator or other flow-restricting device such as a low-flow shower head. Turn the faucet full on and measure the time needed to fill a graduated bucket or other container. Divide gallons by minutes and record the result. Concentrate your efforts to improve flow at points where rates fall short of the figures in the chart.

VALVES AND HOW TO CLEAN THEM

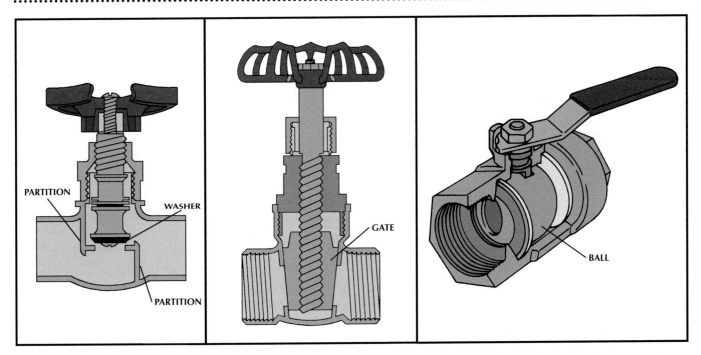

Three common valves.

Globe valves *(above, left)* can be identified by a bulge at the base. They contain a rubber disk that presses against a seat in order to block the flow of water. Although useful as shutoffs to fixtures and branch lines, they contain partitions that impede the flow of water, making globe valves unsuit-

able as main shutoffs. A gate valve *(above, center)* has a gate that raises or lowers as the handle is turned. This unrestrictive design often serves as a main shutoff. Ball valves *(above, right)* fully open or close with a quarter-turn of the handle, which is an advantage if a line must be closed quickly. Like gate valves, ball valves provide an unrestrict-

ed flow and are therefore appropriate for main shutoff valves.

When installing a main shutoff, obtain a valve equipped with a waste, a small drain that can be opened to empty the supply line for plumbing repairs.

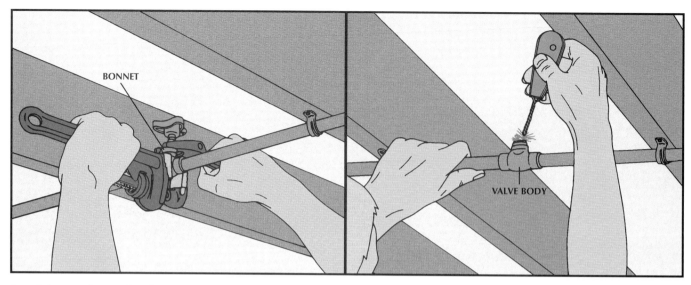

BONNET

VALVE BODY

Servicing a clogged valve.

Ball valves *(page 45)* are unlikely to need cleaning; the ball scrapes off deposits as it turns. Clean a globe or gate valve as shown here:

◆ Close the main shutoff valve and drain the supply system *(page 16)*.

◆ Grip the valve body with a pipe wrench to relieve strain on pipes and loosen the bonnet with an adjustable wrench *(above, left)*. Remove the bonnet and valve stem.

◆ Scour the valve interior with a wire brush designed for cleaning copper fittings *(above, right)*. Bend the shaft of the brush to reach deep inside the valve body.

◆ Reassemble the valve, then shut off the faucets that were opened to drain the system and turn on the water.

REPLACING A VALVE TO IMPROVE FLOW

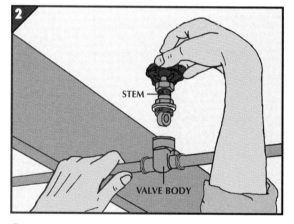

STEM

VALVE BODY

1. Removing the old valve.

◆ Close the main shutoff and drain the system *(page 16)*. If the valve you are replacing is the main shutoff, turn off the water either at the well pump or at the curb valve *(page 11)*.

◆ For copper supply pipe, shown here, open nearby valves on the line to protect their internal components from heat, and grip the adjacent pipe with a rag. Heat one of the soldered joints with a propane torch *(above)*, then pull the pipe out of the valve body, taking care not to bend the pipe. If necessary, turn off the torch and use pliers to pull apart the heated joint before it cools *(inset)*. Repeat for the other joint.

In the case of CPVC pipe, cut the pipe on each side of the valve *(page 23)*. For threaded steel pipe, cut the pipe on one side of the valve; use wrenches to unthread the two cut sections to expose the nearest pipe unions *(page 20)*.

2. Attaching the new valve.

Copper pipe: Solder the valve in place *(pages 17-18)*. Open the valve or remove the stem before soldering to protect the washer *(above)*. When the work cools, insert the stem and turn on the water.

Steel pipe: Adapt the procedure on pages 20 to 22: Attach adapters to the unions and valve inlets and extend CPVC from the inlets to the unions.

After the CPVC cement cures, open the valve and turn on the water.

CPVC pipe: Add threaded adapters to the valve inlets. Cement a short spacer to one adapter; slide a coupling over the spacer. Cement the other adapter to the cut supply line, then join the spacer to the supply line with the coupling. Wait for the cement to cure, open the valve, and turn on the water.

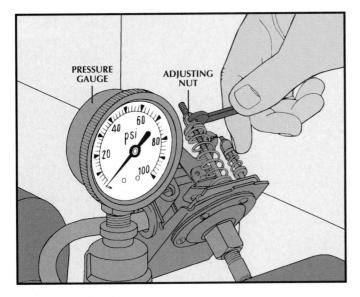

Resetting a pressure switch.

Inadequate flow from a well may result from a low pressure setting. If the pump and pressure tank are rated for a pressure higher than the pressure gauge indicates, try raising the pressure switch setting 5 pounds per square inch. Maximum ratings are usually printed on the components; if they are not, get the figures from the distributor or manufacturer.

◆ Turn off power to the pump.

◆ Remove the cover from the pressure switch, which is located near the pressure gauge *(page 44)*.

◆ With an open-ended wrench, turn the nut atop the taller of the adjustment springs two full turns clockwise *(left)*.

◆ Replace the cover and restore power.

◆ Observe the pump through one operating cycle to see that the pressure does not exceed the desired level.

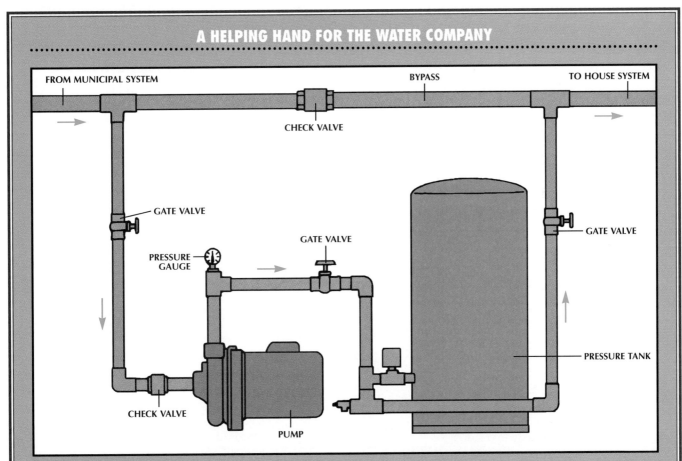

A HELPING HAND FOR THE WATER COMPANY

Spliced into the supply line, a pump-and-pressure-tank arrangement like the one shown here solves the problem of chronically low pressure in a municipal water system. The pump delivers water from the supply line to a T that connects to the tank inlet. The tank stores the water under pressure. When you turn on a faucet or other fixture, water rushes from the tank through the inlet T and into the house system. The resulting drop in tank pressure trips a switch that activates the pump, which operates until pressure is restored. Gate valves in the system can be closed to bypass the pump and tank for maintenance or repair. Check valves, which allow water to flow in only one direction, guard against water flowing backward into the municipal system.

Silencing Noisy Pipes

Pipes can vibrate loudly—and annoyingly—if anchored too loosely to the house framing. More often, however, shrieking or banging sounds coming from your plumbing are a result of water-pressure problems.

Silencing Screeches: In houses close to a water tower or pumping station, abnormally high water pressure is common. It causes bubbling in the pipes, which produces squeals or groans and, over time, can erode valve seats, break apart joints, and make faucets leak.

To confirm that the pressure is too high, measure it close to where the supply line enters the house *(page 44)*. A reading higher than 60 pounds per square inch calls for a pressure-reducing valve near the main shutoff *(below)*.

Cures for Banging Pipes: A phenomenon called water hammer is responsible for the percussive sounds that are often heard when a valve closes abruptly. This sudden halt to the flow of water causes a momentary pressure pulse that produces a loud bang when the rush of water collides forcibly with the closed valve.

Excessive water pressure can cause water hammer, but if pressure is normal, you can correct the problem with a shock absorber or an air chamber installed close to the fixture in question *(opposite)*. Shock absorbers cost more but require no maintenance; air chambers must occasionally be drained.

 **TOOLS**

Pipe wrench
Propane torch
Flameproof pad

 **MATERIALS**

Solder
Flux
CPVC primer and
   cement
Threaded adapters

Supply pipe, fittings
Drain cock
Shutoff valve
Plumbing-sealant
   tape

 **SAFETY TIP**

Wear gloves and safety goggles when soldering copper pipe joints.

A DEVICE FOR LOWERING WATER PRESSURE

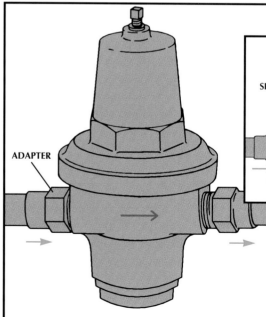

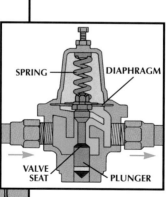

A pressure-reducing valve.

Added to the main supply line *(left)*, this device automatically lowers to a preset level the pressure of water entering the house.

Water on the house side of the valve *(inset, light blue)* presses against a spring-loaded diaphragm to regulate pressure. When the pressure exceeds the value set with an adjustment screw atop the device, the diaphragm bulges upward to lift a plunger toward the valve seat, partially closing the valve and reducing pressure. To install such a valve:

◆ Close the main shutoff and drain the system *(page 16)*.
◆ In a horizontal section of the main supply line, measure and cut *(pages 14-15)* a gap large enough to accommodate the valve and adapters suited to the type of pipe; for a steel pipe, cut at the desired point and unthread the cut sections *(page 20)*.
◆ Secure the valve in place as described on page 46, positioning it so that the arrow on the body points in the direction of water flow.
◆ Turn on the water.

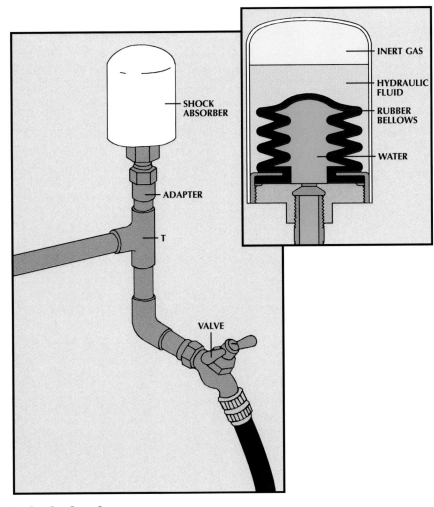

INERT GAS

HYDRAULIC
FLUID

RUBBER
BELLOWS

WATER

SHOCK
ABSORBER

ADAPTER

T

VALVE

CAP

REDUCER

T

DRAIN
COCK

SHUTOFF
VALVE

T

SUPPLY LINE

FIXTURE
VALVE

A shock absorber.

Placed near a valve that causes water hammer
(above), a shock absorber prevents this annoy-
ance with a bubble of gas that cushions the pressure pulse caused by closing a valve *(inset)*. The
bubble is isolated from the water in the pipe carrying the pulse by a rubber bellows surrounded
by hydraulic fluid. To install a shock absorber:
◆ Close the main shutoff valve and drain the system *(page 16)*.
◆ Fit a T to a horizontal section of the pipe sup-
plying the valve.
◆ In the top of the T, place a short pipe and
threaded adapter that accommodates the shock
absorber. Use plumbing-sealant tape to make the
threads watertight, then screw on the shock absorber and tighten it with a pipe wrench.
◆ Connect the fixture to the bottom of the T.
◆ Turn on the water.

Making an air chamber.

In this alternative to a shock absorber, a column of air cushions pressure pulses that cause water hammer. To assemble
an air chamber:
◆ Close the main shutoff and drain the system *(page 16)*.
◆ Install a T as for a shock absorber.
◆ To the upper opening of the T, add a valve and a faucet
to serve as a drain as shown above.
◆ Top off this assembly with an air chamber consisting of a
pipe having at least twice the diameter of the supply pipe,
capped at one end. (For best results, make the air chamber
about 2 feet long.) Connect the air chamber to the supply
line with a reducer.
◆ Turn on the water.

Over time, the chamber may partially fill with water, reducing its effectiveness. Empty it by closing the shutoff and
opening the drain cock.

A typical household consumes up to 150 gallons of water a day. That amount can be cut a third or more by conserving water *(checklist, right)* and by modifying plumbing fixtures to reduce the amount of water they discharge.

Indoor Fixtures: Installing a water-saving aerator makes a kitchen faucet more efficient. In a shower head, a flow restricter *(below)* reduces the stream from 5 gallons a minute to as little as 2 gallons a minute while delivering an adequate spray. Lowering the volume of water in an old-style large toilet tank can cut the amount of each flush in half *(opposite, top)*.

Water and fuel are saved by insulating exposed hot-water pipes *(page 96)* so that the water does not lose heat as fast as it would in uninsulated pipes. Kitchen hot-water dispensers and refrigerated drinking water also eliminate losses that occur while waiting for the water to become hot or cold.

Regular Maintenance: An obviously dripping faucet or running toilet is easy to locate, but water often seeps through worn fixtures without obvious signals. Check faucets, spigots, and toilets for leaks throughout the house at least twice a year, and repair them with the techniques shown in Chapter 3.

Outdoor Conservation: Make the most of water applied to lawns and gardens, especially in hot, dry weather. A sprinkler timer prevents wasteful overwatering. Some timers must be set for each watering; others, like the one shown at the bottom of the opposite page, are programmable for longer periods.

Spreading a layer of mulch on garden beds and cutting grass high reduces evaporation, as does watering in the evening.

Water Recycling: A gray-water reclamation system like the one illustrated on pages 124 to 125 filters water drained from sinks, tubs, showers, and washing machines. The water can be used to irrigate gardens and lawns or, in some states, to flush toilets.

TOOLS

Wire brush
Pipe wrench

MATERIALS

Plumbing-sealant   Water dam
     tape   Food coloring

Water-Saving Tips

Kitchen
✔ Soak dirty dishes and pans instead of rinsing them with running water.
✔ Turn on the dishwasher only when it is full; select a water-saving cycle.
✔ Minimize use of the garbage disposer; compost vegetable waste instead.

Bathroom
✔ Take a short shower instead of a bath, since filling a tub requires more water; turn water off while soaping and shampooing, on again for rinsing.
✔ Run water intermittently while washing your hands and face, brushing your teeth, or shaving.
✔ Flush the toilet only for human waste.
✔ Adjust the flush mechanism to a lower water level.

Laundry
✔ Wash full loads whenever it is possible.
✔ Lower fill level for small loads.
✔ Presoak heavily soiled clothes.

Lawn and garden
✔ Select plant varieties that need little water.
✔ Consider watering with a soaker hose instead of a sprinkler, and turn it on in the evening to minimize evaporation.
✔ Cover swimming pools to save 50 gallons of water on a sunny day.

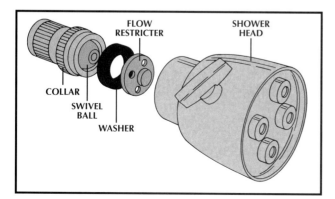

A shower-head flow restricter.
◆ Remove the shower head from the shower arm *(page 68)*.
◆ Unscrew the collar from the shower-head assembly and insert a flow restricter as illustrated at left. Reassemble the shower head.
◆ Clean the shower-arm threads with a wire brush, wrap them with plumbing-sealant tape, then reinstall the shower head.

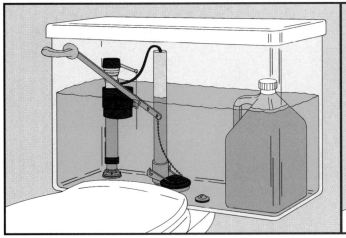
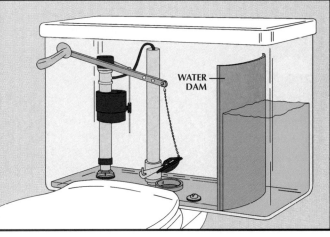

Reducing toilet-tank volume.
Place a water-filled plastic bottle in the tank *(above, left)*, or install a self-sealing water dam, a plastic sheet that retains some of the tank water during a flush *(above, right)*. Do not allow either addition to interfere with the flushing mechanism.

TRICKS OF THE TRADE

Detecting Toilet Leaks

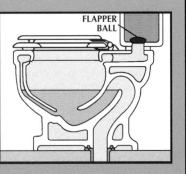

Often invisible and inaudible, a tank-ball or flapper-ball leak can be a big water waster. To check your toilet for such a leak, pour food coloring into the tank and wait 20 minutes. If water in the bowl becomes tinted, replace the old tank ball or flapper ball *(page 83)*.

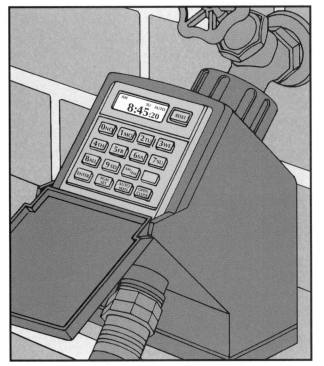

A lawn-watering timer.
The battery-powered device shown at left contains an electrically operated valve controlled by a timer that can be programmed for a weekly cycle of lawn watering. Install such a device as follows:
◆ Wrap the threads of the sillcock and the timer with plumbing-sealant tape.
◆ Screw the timer to the sillcock and the sprinkler hose to the timer.
◆ Program the timer with the instructions provided, then open the sillcock.

Whether it comes from a municipal water company or a private well, your household water may contain impurities. Fortunately, a wide range of water-treatment equipment is available to homeowners, and each device may be used with others to cure almost any combination of conditions.

Testing the Water: The first column of the chart on the opposite page describes the signs of possible water-supply problems. If any of these indicators are present, your water should be tested. You may be able to conduct the investigation yourself with a home test kit *(opposite, bottom)*. Some conditions, however, require professional analysis. Your local board of health can provide a list of water-testing laboratories.

If you suspect that there is lead in your water, have a commercial laboratory run a test. A problem with lead is likeliest in older homes, which may have copper plumbing joined by lead-containing solder or—especially in large cities—may be linked to the water main by lead entry piping.

Since there is no sure way for a homeowner to detect the presence of bacteria, a laboratory test should be made for any new well. An existing well should be tested every 6 months if contamination is known to occur in the area. Your local board of health will do a bacterial analysis at no charge. The extension service of the county or state department of agriculture also may offer free tests.

Hard Water: Calcium and magnesium are responsible for the widespread condition known as hard water; these minerals reduce the efficacy of soap, corrode appliances, and can even block pipes. Treat the problem by installing a water softener—a device that can also remove other impurities. A few regular maintenance procedures will keep a softener running smoothly *(pages 54-56)*.

A Range of Filters: Many styles of filters are available to treat water, and they can be used either alone or in conjunction with a water softener. Install the filter in the line ahead of a water softener to screen out sediment *(page 57)*. Replenish neutralizing, oxidizing, and carbon filters according to the manufacturer's instructions—as often as every 3 days or as infrequently as every 3 years, depending on the type of filter.

Other Water-Treatment Units: A reverse-osmosis purifier forces water through a thin membrane to remove unwanted materials. Some reverse-osmosis units are designed to serve the entire house system. More commonly, these devices have a smaller capacity, and are attached under a sink to provide drinking and cooking water from a single tap. The membrane must be replaced periodically.

Chemical feeders inject a small amount of chemical solution into the water to counteract pollutants. Supply the feeder with chemicals on a regular basis.

TOOLS

Broom	Strap wrench
Wet/dry vacuum	Pipe cutter
Adjustable wrench	Propane torch
Oven baster	Flameproof pad
Screwdriver	
Socket wrench	
Pliers	

MATERIALS

Supply pipe and fittings
Gate valves
Plumbing-sealant tape
Solder
Flux

SAFETY TIPS

Gloves and safety goggles provide protection when soldering pipe.

Common Water Problems and Their Remedies

Problem	Cause	Solution
Low sudsing power of water. Soap deposits on fixtures and clothes. White scale in pipes and water heaters.	Hard water (contains calcium and magnesium)	A water softener.
Rusty deposits in sinks, tubs, and washing machines and on washed fabrics. Water left standing turns reddish.	Iron compounds	If the problem is minor, a water softener. Otherwise, an oxidizing filter.
Green or blue stains in fixtures served by copper pipes. Red stains below faucets in fixtures served by steel pipes.	Corrosion of pipes by acidic water	If the problem is minor, a neutralizing filter. Otherwise, a chemical feeder with an alkaline solution.
Water has a "rotten egg" smell, blackish tinge, tarnishes silver.	Hydrogen sulfide	If the problem is minor, an oxidizing filter. Otherwise, a chemical feeder with a chlorine solution, followed by a sand filter.
Water has a yellow or brownish tinge or an unpleasant taste or odor.	Tannins, algae, humic acid, or other organic substances	If the problem is minor, a carbon filter. Otherwise, a chemical feeder with a chlorine or alum solution, followed by a sand filter or, for potable water only, a reverse-osmosis unit.
Water has an alkaline or soda taste.	Sodium salts	A water softener or oxidizing filter.
Water has a chlorine odor.	Excessive chlorine	A carbon filter or reverse-osmosis unit.
Water has a cloudy or dirty appearance; seats, valves, and moving parts in appliances wear out quickly.	Suspended particles of silt, mud, or sand	A sediment or sand filter.
Intestinal disorders and disease result from drinking the water.	Coliform bacteria from improperly disposed sewage nearby	Correct improper sewage disposal and add a chemical feeder with a chlorine solution, followed by a carbon filter.
Poisoning or disease results from drinking the water.	Lead, herbicides, pesticides, chlorine	A carbon filter.
	Nitrates, chlorides, sulfates, lead, chlorine	A reverse-osmosis unit.

DO-IT-YOURSELF TESTS

With various home test kits, you can easily test water for hardness, pH (the degree of acidity or alkalinity), the presence of minerals or organic substances, and several other conditions. In some tests, paper strips are dipped in water and change color when certain elements are present. Other tests use chemicals dropped into a water sample; again, a change of color indicates impurities.

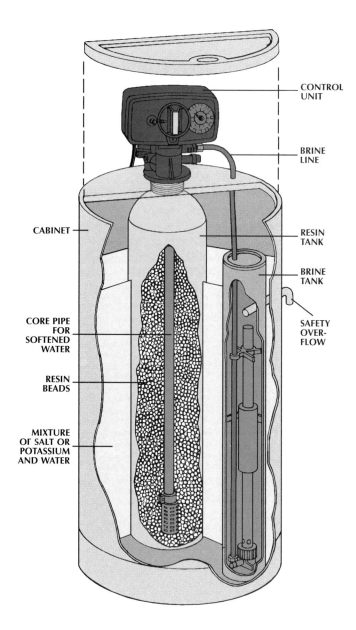

CONTROL UNIT

BRINE LINE

CABINET

RESIN TANK

BRINE TANK

CORE PIPE FOR SOFTENED WATER

SAFETY OVER-FLOW

RESIN BEADS

MIXTURE OF SALT OR POTASSIUM AND WATER

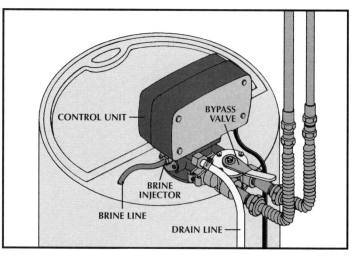

CONTROL UNIT

BYPASS VALVE

BRINE INJECTOR

BRINE LINE

DRAIN LINE

Anatomy of a water softener.

A water softener passes hard water over resin beads in a central tank; the resin contains salt or potassium, which is exchanged for the calcium and magnesium impurities in the water during its transit. Periodically, the beads must be chemically revived and the impurities removed. This is accomplished by a process called backwashing: Brine, formed by mixing water with a supply of salt or potassium outside the resin tank, is washed over the beads, recharging them and also carrying away the accumulated calcium and magnesium. An electrically powered control unit regulates the backwash cycle with either a timer or a meter that keeps track of water use. A bypass valve, used during maintenance tasks, allows water to flow to the household without passing through the softener.

Inspecting the salt or potassium.

Check on the salt or potassium supply once a week.
◆ Remove the lid of the cabinet and look inside.
◆ If a hard crust has formed on the salt or potassium supply, break it into pieces with a broom handle *(right)*.

Accumulated dirt in the cabinet means the salt or potassium contains impurities. Correct the situation as follows:
◆ Wait until the supply is low. Unplug the softener, lift the lid, and empty the cabinet with a wet/dry vacuum.
◆ Clean the cabinet with an abrasive bathroom cleanser. Rinse thoroughly.
◆ Vacuum out the remaining water. To prevent corrosion, rinse out the vacuum before putting it away.
◆ Replace the salt or potassium with a pure product.

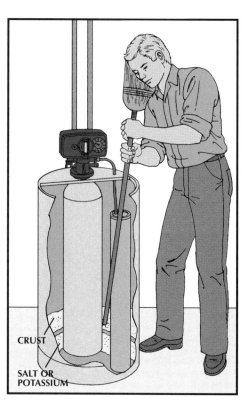

CRUST

SALT OR POTASSIUM

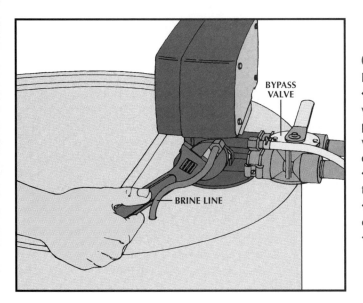

Clearing the brine line.

Every 6 months, check the brine line for blockages.

◆ Unplug the softener, turn the bypass valve to divert the water supply, and set the control-unit dial to the backwash position to relieve pressure on the brine line. On a unit without an integral bypass valve, close the inlet valve and open the nearest water faucet.

◆ Loosen the compression nut connecting the brine line to the control unit and pull the line free *(left)*.

◆ If there is an obstruction in the line or the inlet, use an oven baster to flush it out with warm water.

◆ Reattach the brine line.

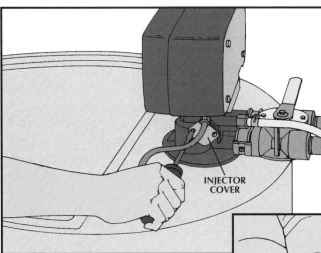

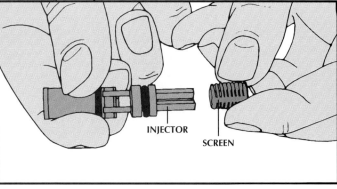

Cleaning the injector.

The injector regulates the flow of brine to the resin tank. Clean it after inspecting the brine line, or at any other time.

◆ With the softener unplugged and in the bypass mode, remove the screws holding the injector cover to the control unit *(above)* and pull the cover off.

◆ Turn the injector counterclockwise by hand or with a socket wrench and then pull it from its housing.

◆ Remove the small filter screen from the injector *(above, right)*.

◆ Clean a clogged screen in warm, soapy water; replace a broken one. Remove any obstruction in the injector by blowing gently. Do not use a sharp object to remove a blockage.

◆ Replace the screen, screw the injector back into its housing, replace the injector cover, and tighten the screws.

◆ Turn the bypass valve to restore the water supply, reset the control-unit dial for normal operation, and plug in the softener. On a unit without an integral bypass valve, close the open water faucet and open the inlet valve.

REPLACING THE SOFTENER CONTROL UNIT

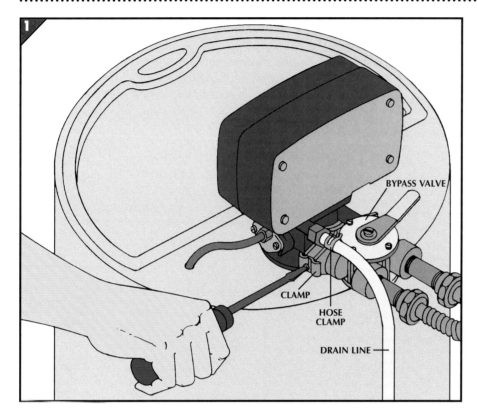

1. Disconnecting the control unit.

◆ Unplug the softener, put it in bypass mode, and set the control-unit dial in the backwash position to relieve pressure on the brine line. On a unit without an integral bypass valve, close the inlet valve and open the nearest water faucet.

◆ Remove the brine line *(page 55)*.

◆ Loosen the screws on the clamps that hold the bypass valve to the control unit *(left)*, then pull the valve from the unit. On a unit without an integral bypass valve, unscrew the fittings joining the inlet and outlet pipes to the control unit and pull out the pipes.

◆ With a pair of pliers, compress the prongs of the hose clamp on the drain line, and pull the line free.

◆ Push the pipes back to provide space to remove the control unit.

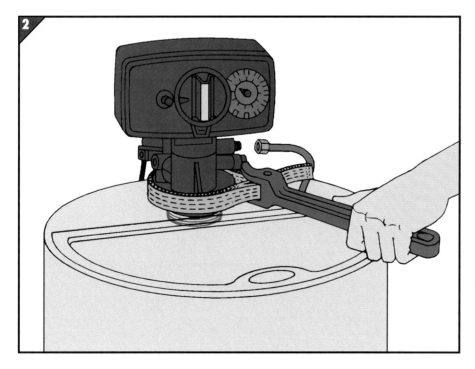

2. Replacing the control unit.

◆ Tighten a strap wrench around the neck of the control unit and turn in a counterclockwise direction until the control unit is free.

◆ Set a new control unit on top of the cabinet and turn it clockwise by hand, then tighten it with the strap wrench.

◆ Replace the bypass valve or the inlet and outlet pipes, and reconnect the brine line and the drain line.

◆ Turn the bypass valve to restore the water supply, reset the control-unit dial for normal operation, and plug in the softener. On a unit without an integral bypass valve, close the open water faucet and open the inlet valve.

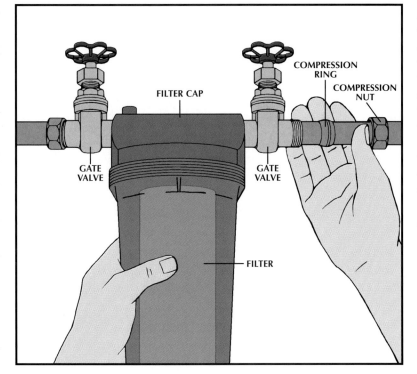

A horizontal line installation.

◆ Turn off the water at the main shutoff valve and drain the supply pipe *(page 16)*.

◆ Using plumbing-sealant tape, thread a gate valve *(page 45)* with exterior threads onto each side of the filter cap and tighten them until each valve is in an upright position.

◆ Measure and cut out a section of the supply pipe long enough to accommodate the filter and valves *(page 14)*.

◆ Slide a compression nut and compression ring over each cut pipe end, fit the valves over the pipe ends, and tighten the nut and ring onto each valve.

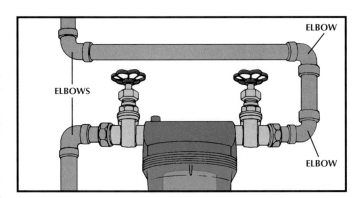

A vertical line installation.

◆ Turn off the water at the main shut-off valve and drain the supply pipe *(page 16)*.

◆ Cut out a 4-inch section of the pipe and attach elbows to each cut end.

◆ Create a loop, installing the filter in the lower leg of the loop with gate valves, as shown at left.

◆ Attach the loop to the elbows on the supply pipe.

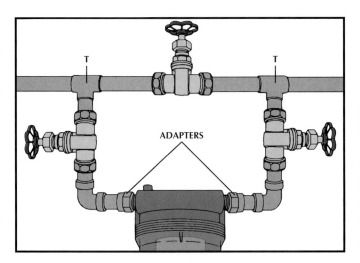

Creating a bypass loop.

◆ Turn off the water at the main shutoff valve and drain the supply pipe *(page 16)*.

◆ Create a bypass loop, using compression fittings, gate valves, and elbows.

◆ Attach the filter to the loop with two threaded adapters wrapped with plumbing-sealant tape.

◆ Cut out a section of the supply line long enough to accommodate the bypass loop *(page 14)*, insert a T fitting at each end of the pipe, and install a gate valve between the two T fittings.

◆ Attach the loop to the T fittings on the supply pipe.

Showers, Sinks, and Toilets

The most visible parts of any plumbing system are the sinks, toilets, showers, and other devices such as the water heater that stand ready to take up water from the house supply lines and put it to use. With the techniques in this chapter, you can maintain and make basic repairs to all of this equipment, as well as replace items that are worn out or old-fashioned.

Working on Tub and Shower Fittings

Three-Handle Tub Fixtures
Single-Lever Ball Faucets
Single-Lever Cartridge Faucets
Replacing a Three-Handle Fixture
Installing a New Single-Lever Faucet
Servicing and Replacing Shower Diverters
Unclogging a Shower Head
Adjusting Bathtub Drains

Maintaining and Replacing Sink Fittings

Servicing Aerators and Spray Attachments
Curing Seepage around a Kitchen Strainer
Solving Garbage Disposer Leaks
A Remedy for a Leaking Washbasin Drain
Double-Handle Faucets
Repairs for Single-Lever Cartridge Faucets
Servicing Single-Lever Ball Faucets
Fixing Single-Lever Ceramic-Disk Faucets
Replacing a Faucet

Toilet Repair and Installation

Restoring a Toilet to Faultless Operation
Stopping Leaks
Replacing the Fill Mechanism
Servicing a Pressure-Flush Toilet
Putting in a Modern Toilet

The Ins and Outs of Water Heaters

Lighting a Gas Burner
Small Jobs That Make a Difference
Replacing the Thermocouple
Dealing with a Faulty Relief Valve
Servicing the Drain Valve
Testing Electric Heating Elements
Installing an Electric Water Heater

Removing a shower head →

Bathtub and shower faucets and other fittings may develop a variety of troubles—among them, faucet leaks, erratic shower flow, the seeping away of water during a bath, or the refusal of a diverter to send water to a shower head. Often, the problem can be dealt with by simple servicing and repairs.

Easy Fixes: Begin a faucet repair by turning off the water supply and opening the faucet to empty the lines. Close the drain and cover it with padding or a large cloth to pro-

tect the tub or shower and prevent loss of parts. As you take the faucet apart, line up the parts in sequence to facilitate reassembly.

When fittings of any kind are removed in order to be repaired, clean away mineral deposits with white vinegar and an old toothbrush. Replace any components that appear worn, pitted, or corroded. A replacement part must be a duplicate of the original; to ensure that the match is exact, take the old part to the plumbing-supply store when you purchase a substitute.

All-New Hardware: Some problems can be solved only by replacing the entire fitting. The biggest such job is the installation of a new faucet set, which is required if the old faucet body (the portion that is directly connected to the pipes) is severely worn or corroded. This project calls for the basic pipe fitting skills described on pages 16 to 25, and it may also present challenges of access. If it is necessary for you to cut through a wall to get at the pipes, create an access panel as part of the repair *(page 64)*.

TOOLS

Screwdrivers
Faucet-handle
 puller
Hex wrenches
Ratchet wrench
 with deep socket
Tube cutter with
 reamer

Pliers
Propane torch
Pipe wrench
Hammer and chisel
Dry-wall saw
Minihacksaw
Old toothbrush

MATERIALS

White vinegar
Padding or a large cloth
Penetrating oil
Silicone lubricant
Kit of faucet-replacement
 parts
1 x 2 wood cleats
Wood screws

Replacement pipe and
 fittings
Solder
Flux
CPVC primer and cement
Pipe clamps
Flameproof pad
Silicone sealant

Plumbing-sealant tape
Masking tape
Steel wool
Small wire

SAFETY TIPS

Wear gloves and safety goggles when soldering copper pipe joints.

THREE-HANDLE TUB FIXTURES

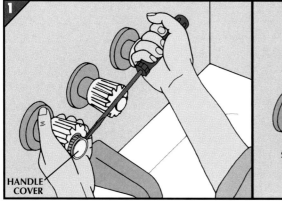

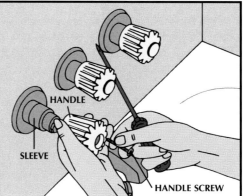

HANDLE
COVER
HANDLE
SLEEVE
HANDLE SCREW

1. Removing the handle.
◆ Pry off the handle cover with a screwdriver *(far left)*.
◆ Remove the screw, handle, and sleeve *(near left)*. If the handle is stubborn, pour hot water over it to cause expansion, then pry it off at the base with a screwdriver or lift it off with a faucet-handle puller *(page 9)*.

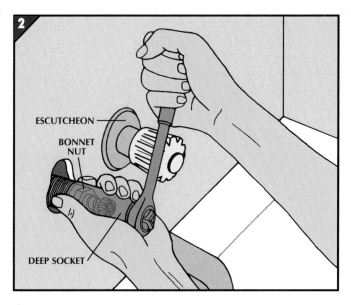

2. Removing the stem.

◆ If the escutcheon is fastened by a setscrew, loosen it with a hex wrench. Pry the escutcheon from the wall.

◆ If the bonnet nut on the stem is beneath the wall surface, chip away any obstructing tile or grout.

◆ With a ratchet wrench and a deep socket, remove the bonnet nut and stem, turning the ratchet counterclockwise.

◆ If the stem sticks in the faucet, apply penetrating oil and wait 15 minutes before trying again.

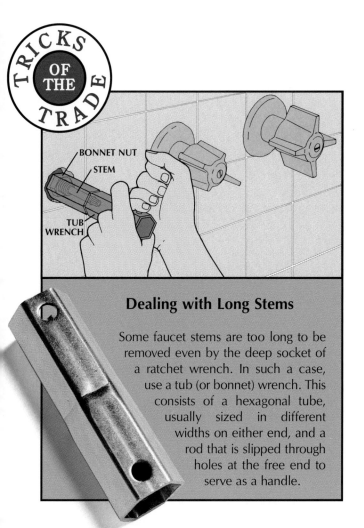

Dealing with Long Stems

Some faucet stems are too long to be removed even by the deep socket of a ratchet wrench. In such a case, use a tub (or bonnet) wrench. This consists of a hexagonal tube, usually sized in different widths on either end, and a rod that is slipped through holes at the free end to serve as a handle.

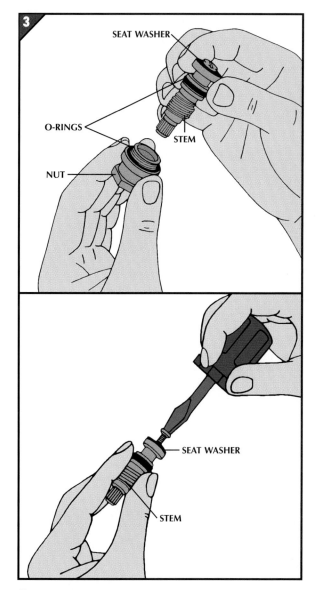

3. Curing leaks.

◆ Separate the stem from the bonnet nut (top picture). If they do not unscrew easily, reattach the faucet handle for better leverage.

◆ If the faucet is leaking around the handle, pry off and replace the O-rings. Replace the entire stem assembly if it appears worn.

◆ For leaks from the tub spout or shower head, remove the screw that holds the seat washer to the stem (bottom picture) and replace the old washer with a new one of the same size.

◆ Inspect the valve seat in the wall. If it is smooth and shiny, apply a thin layer of silicone lubricant. If it appears damaged, replace it (page 62).

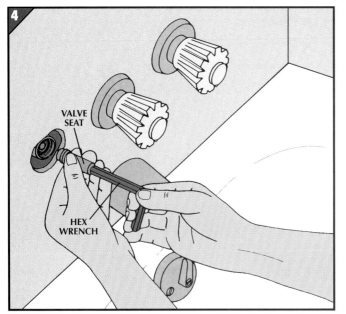

4. Replacing the seat.
◆ To remove a worn valve seat, insert a large hex wrench into the faucet body and turn it counterclockwise.
◆ Buy an exact replacement of the valve seat, lubricate the outside, and screw it into the faucet body.

SINGLE-LEVER BALL FAUCETS

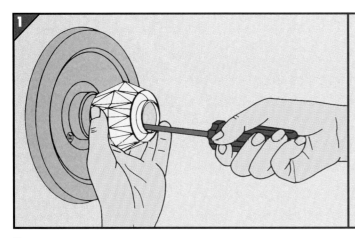
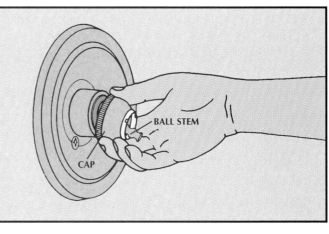

1. Removing handle, cap, and ball.
To cure leaks from the tub spout or shower head, buy a kit of replacement parts for the faucet make and model, and replace the seats, springs, and ball.
◆ Pry off the handle cover with a screwdriver.

◆ Remove the handle screw *(above, left)* and then lift off the handle.
◆ Unscrew and lift off the cap *(above, right)*. Pull the ball stem to lift off the ball-and-cam assembly that controls the mixture of hot and cold water.

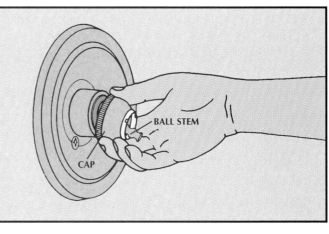

2. Replacing the seats and springs.
◆ With a small screwdriver or nail, lift the rubber seats and springs out of the two small sockets in the faucet body *(left)*.
◆ Set in new springs. If the springs are cone shaped, insert the large end first.
◆ Place new seats over the springs.
◆ Set the ball in the faucet body, adjust the cams over the ball, and screw on the cap.

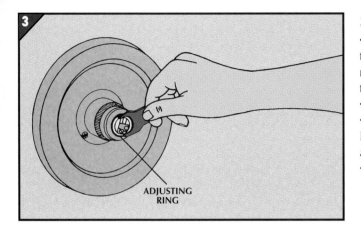

3. Tightening the adjusting ring.

◆ With pliers or a special wrench often included in a ball-type replacement kit, turn the adjusting ring inside the cap clockwise to tighten it *(left)*.
◆ Turn on the water supply.
◆ Open the faucet with the ball stem. If water leaks around the stem, install a new ball-and-cam assembly as well.
◆ Replace the handle.

ADJUSTING RING

SINGLE-LEVER CARTRIDGE FAUCETS

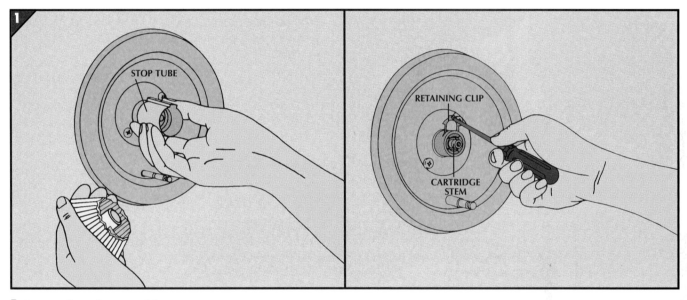

STOP TUBE

RETAINING CLIP

CARTRIDGE STEM

1. Removing the cartridge.

◆ Remove the handle as explained for a ball-type faucet *(opposite, Step 1)*.
◆ Pull off the stop tube *(above, left)* to reveal the stem of the cartridge and its retaining clip, which secures the cartridge to the faucet body. (On some faucets, the escutcheon must first be removed.)
◆ Pry up the cartridge retaining clip with a nail or small screwdriver *(above, right)*. Slide the clip all the way out with a screwdriver or pull it out with long-nose pliers.

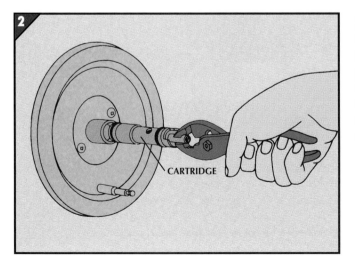

CARTRIDGE

2. Replacing the cartridge.

◆ Grip the cartridge stem with pliers and, rotating the stem first one way and then the other, pull it free from the faucet body *(left)*.
◆ Insert the new cartridge with the flat part of the stem facing up. (Otherwise, the hot and cold water will be reversed.)
◆ Reinstall retaining clip, stop tube, and handle.

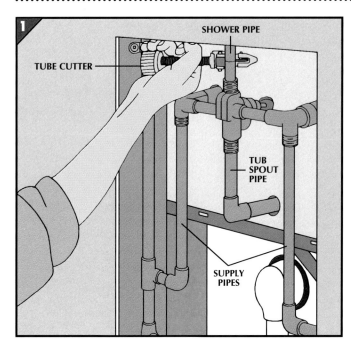

TUBE CUTTER

SHOWER PIPE

TUB SPOUT PIPE

SUPPLY PIPES

1. Cutting the pipes.

◆ Detach all handles, escutcheons, and the tub spout from the old faucet body.

◆ Look for an access panel on the other side of the wall. If there is no panel, cut through the wall, as explained below.

◆ For copper or CPVC supply pipes, cut the pipes leading into the faucet body *(left)* at points where they can easily be reconnected with fittings to the new faucet. If the supply lines are threaded steel, cut them and unthread the cut sections *(page 20)*.

◆ Take the faucet body, with its cut pipe ends still in place, to a plumbing-supply store. Buy a replacement, fittings, and spacer pipes.

INSTALLING AN ACCESS PANEL

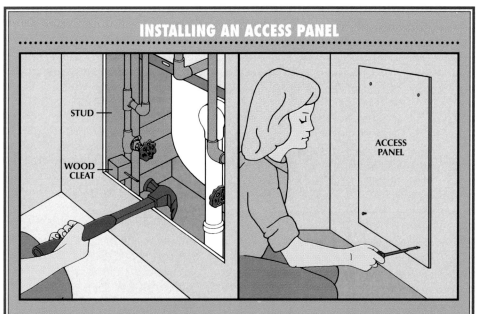

STUD

WOOD CLEAT

ACCESS PANEL

If a faucet body is walled in, create an access panel as part of the replacement job. To determine where to cut, measure from the floor and nearest corner to the fitting on the tub side of the wall; transfer these measurements to the other side of the wall, and punch a hole with a hammer and chisel at the indicated point. With a dry-wall saw, cut from the hole out to the studs. Draw a rectangular box, stud to stud and about 2 feet high, and cut the wall along the outline. Cut a plywood panel slightly larger than the opening. At the corners of the opening, nail wood cleats to the studs *(above, left)*. The panel attaches to the cleats with screws *(above, right)*.

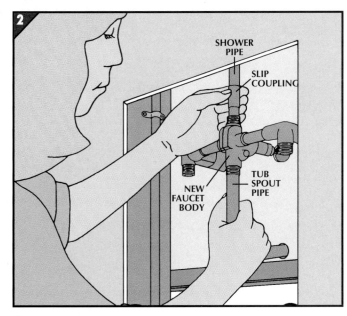

2. Connecting the shower pipe.

◆ Cut the pipes to connect the new faucet body to the shower pipe, the spout, and the supply lines. For CPVC or steel supply lines, use CPVC *(pages 20-23)* and add threaded adapters to the faucet body.

◆ Temporarily place the pipes in the faucet body, and put the assembly in position to check the pipes' lengths. Remove it and shorten them as needed.

◆ Secure the tub spout pipe and the shower pipe spacer to the faucet body, placing a slip coupling on the spacer.

◆ Set the faucet body assembly in place.

◆ Add an elbow and horizontal pipe to the tub spout pipe. Use a slip coupling to attach the shower pipe to the spacer *(above)*.

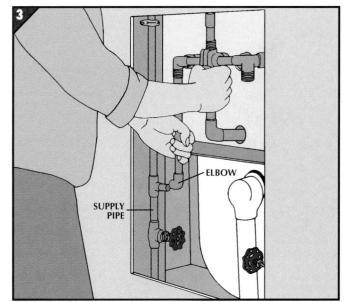

3. Connecting the supply pipes.

◆ Slip a supply pipe into one of the two openings in the faucet body.

◆ At the other end of the pipe, attach a coupling or elbow and connect it to the supply line *(above)*. Connect the other pipe in the same way.

◆ Solder or cement the pipes at both ends (see pages 17 and 18 for copper, pages 22 and 23 for CPVC or steel).

◆ Leave the access panel open for a few days in order to check for leaks.

ANTISCALD PROTECTION

Many state codes require so-called pressure-balancing faucets in new installations to protect against dangerous changes in the mix of cold and hot water—as when a toilet is flushed while someone is showering. These faucets, available only in single-lever style, have a mechanism that regulates the relative pressure of the hot- and cold-water supplies—hence the maximum temperature. With two- or three-handle faucets you can install a device such as the one at right—a separate pressure-balancing valve. This may require rerouting the existing supply pipes, since the valve's internal mechanism is sandwiched between two parallel intakes and outflows.

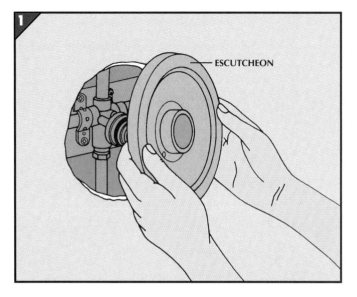

1. Gaining access to the faucet body.
◆ Remove the handle *(page 62)* and unscrew and lift off the escutcheon *(left)*.
◆ If you have access to the faucet from the other side of the wall, follow the procedure for removing the faucet body described on page 64 for a three-handle faucet. If you cannot gain access from the back, you may be able to replace the faucet body from the tub side *(next step)*; do not attempt this, however, in a house with steel supply lines.

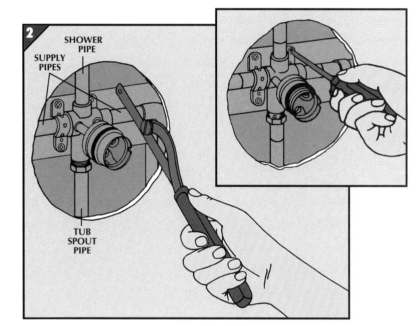

2. Removing the faucet from the tub side.
◆ Cut the shower pipe, tub spout pipe, and supply pipes with a minihacksaw *(left)*. It may be necessary for you to chip away tiles or wallboard for easier access.
◆ Remove the screws that secure the faucet body to the wooden crosspiece behind it *(inset)*.
◆ Lift the faucet body out of the wall.
◆ Buy a replacement faucet of the same make and model as the old one (or at least the same size), along with replacement pipe and slip couplings that will be used to reconnect the pipes. To calculate the total amount of new pipe needed, add the lengths of the four pipes in the old faucet plus at least half an inch for each length.

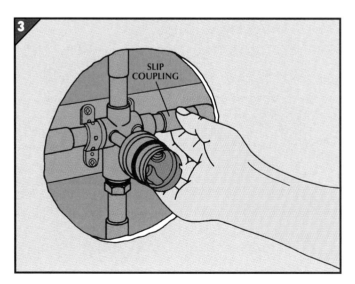

3. Reconnecting the pipes.
◆ Cut the new pipe into segments that will join the faucet body to the shower pipe, tub spout pipe, and supply lines; insert them in the faucet body, using threaded adapters for plastic pipe. (See pages 17 and 18 for working with copper pipe, page 23 for working with plastic.)
◆ Position the faucet and connect the pipes with slip couplings *(left)*.
◆ Secure the faucet body by screwing it to the crosspiece with pipe clamps.
◆ If you have copper pipe, remove all plastic and rubber parts from the faucet body before soldering the joints. Mount a flameproof pad between the crosspiece and the pipes for protection. (You may need to unfasten the clamps to slip the cloth through.)
◆ Screw on the escutcheon and reconnect the handle.

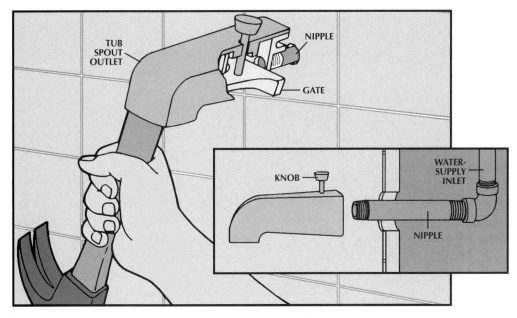

Replacing a screw-on spout diverter.

The knob of this diverter raises an internal gate that closes the pathway to the tub spout, forcing the water up to the shower head. If the mechanism malfunctions, the entire spout must be replaced.

◆ To remove the old spout, insert a piece of wood, such as a hammer handle, in the spout and turn it counterclockwise. Buy a replacement spout of the same length as the old one.

◆ If you cannot match the spout, buy an appropriately sized threaded adapter or replace the nipple *(inset)*.

◆ Apply plumbing-sealant tape to the threads and silicone sealant to the spout base.

◆ Hand tighten the spout. If you cannot complete the alignment by hand, use the makeshift wood tool or—provided the spout comes with a pad to protect the finish—use a wrench.

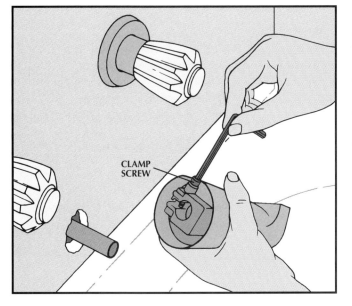

Replacing a slip-fit spout diverter.

◆ To remove this type of spout, loosen the clamp screw on the underside with a hex wrench. Grasp the spout firmly and twist it off the pipe.

◆ With the hex wrench, loosen the clamp screw on the new spout *(left)*, and twist the spout onto the pipe.

◆ Turn the spout so that the clamp screw faces up, and partially tighten the screw.

◆ Twist the spout into position and finish tightening.

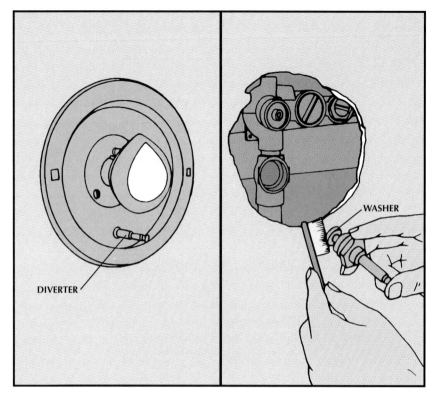

Servicing and replacing other diverter types.

For a diverter in a single-lever faucet unit *(above, left),* remove the faucet handle and escutcheon, then unscrew the diverter.

◆ If water is not being diverted from the tub spout to the shower head properly, clean any sediment that may be adhering to the washer with white vinegar and an old toothbrush *(above, right).*

◆ If water leaks from around the diverter or if its parts are worn, replace the entire mechanism with one of the same make.

◆ For a diverter in the center of a three-handle faucet, replace the entire cartridge if it leaks or malfunctions. Proceed just as you would for a faucet cartridge *(page 63).*

UNCLOGGING A SHOWER HEAD

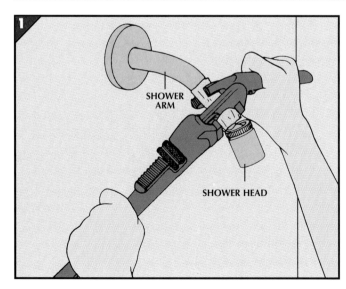

1. Removing a shower head.

When the water flow from a shower head is uneven or insufficient, disassemble and clean the fitting:

◆ Wrap the shower head collar in masking tape and turn it counterclockwise with a pipe wrench. For greater leverage, grip the shower arm with one wrench and turn the collar with a second wrench *(left).*

◆ Twist off the loosened shower head by hand.

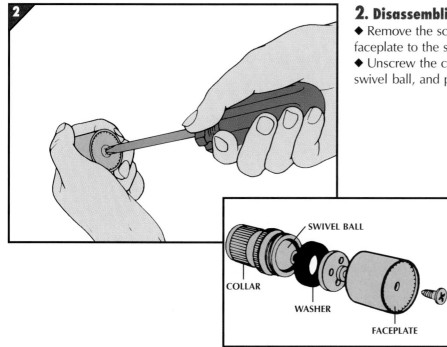

2. Disassembling the shower head.
◆ Remove the screw *(left)* or the knob that secures the faceplate to the shower head.
◆ Unscrew the collar from the shower head to reveal the swivel ball, and pry out the washer.

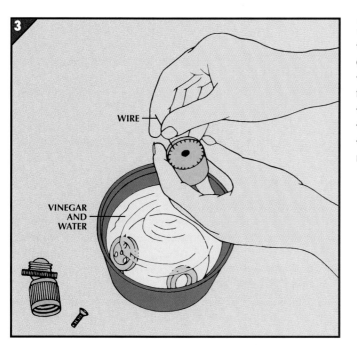

3. Cleaning the shower head.
◆ Soak the entire shower head or its disassembled parts overnight in a bowl of white vinegar and water.
◆ Scrub with steel wool or an old toothbrush, and clear the spray holes with a small wire *(left)*.
◆ Rinse thoroughly.
◆ Replace any worn parts of the shower head.
◆ Lubricate the swivel ball with silicone lubricant, and reassemble the head.

A RANGE OF VERSATILE SHOWER HEADS

The traditional shower head, designed simply to deliver a steady stream of water from above, is only one of many shower fittings now available. Today's alternatives include hardware with internal mechanisms that conserve water by lowering the pressure, reduce or shut off water when the temperature gets too high, or regulate the pulse of the spray to simulate a massage. Hand-held showers, attached either to the tub spout or the shower arm, allow you to maneuver the spray freely—and you can also hang the head from a bar and use it in the conventional manner.

ADJUSTING BATHTUB DRAINS

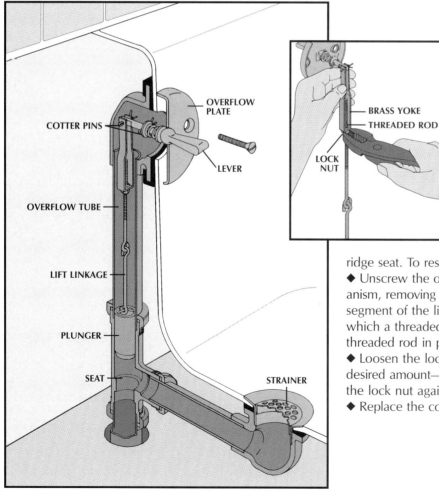

OVERFLOW PLATE

COTTER PINS

LEVER

OVERFLOW TUBE

LIFT LINKAGE

PLUNGER

SEAT

STRAINER

BRASS YOKE

THREADED ROD

LOCK NUT

A trip-lever drain.

The key element of this type of drain is a brass plunger suspended from a lift linkage. The lever lowers the plunger onto a slight ridge below the juncture of the overflow tube and the drain, blocking outflow of water via the main tube outlet; however, any water that spills into the overflow tube can pass freely down the drain because the plunger is hollow. A leak in the drain may be due to wear on the plunger caused by repeated impact against its ridge seat. To restore a proper fit, lengthen the linkage:

◆ Unscrew the overflow plate and lift out the whole mechanism, removing any accumulated hair or debris. The upper segment of the lift linkage consists of a brass yoke from which a threaded rod is suspended; a lock nut secures the threaded rod in place.

◆ Loosen the lock nut with pliers, turn the threaded rod the desired amount—usually just a turn or two—then tighten the lock nut again *(inset)*.

◆ Replace the cotter pins if they are corroded.

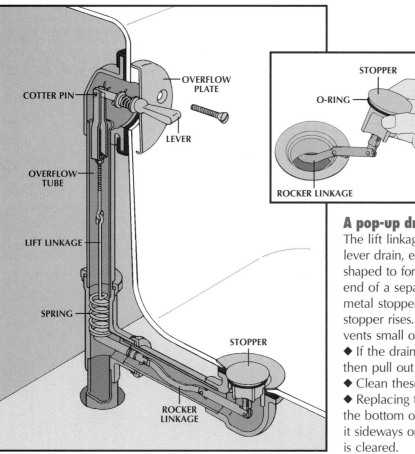

COTTER PIN

OVERFLOW PLATE

LEVER

OVERFLOW TUBE

LIFT LINKAGE

SPRING

STOPPER

ROCKER LINKAGE

STOPPER

O-RING

ROCKER LINKAGE

A pop-up drain.

The lift linkage of a pop-up drain resembles that of a trip-lever drain, except that the lower end of the linkage is shaped to form a stiff spiral spring. This spring rests on the end of a separate, horizontal rocker linkage leading to the metal stopper. When the spring presses downward, the stopper rises. The stopper has a cross-shaped base that prevents small objects from passing down the drain.

◆ If the drain begins to clog, open it with the control lever, then pull out the stopper and the rocker linkage.

◆ Clean these parts of accumulated hair.

◆ Replacing the stopper can be tricky. Make sure that the bottom of the curve in the linkage faces down. Work it sideways or back and forth until the bend in the pipe is cleared.

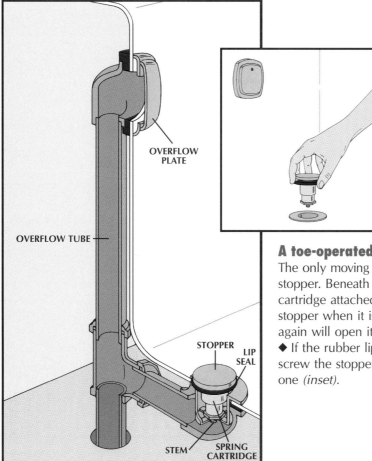

OVERFLOW PLATE

OVERFLOW TUBE

STOPPER

LIP SEAL

STEM

SPRING CARTRIDGE

A toe-operated drain.

The only moving part of this type of drain is the stopper. Beneath its rubber lip seal is a spring cartridge attached to a stem. Pressing on the stopper when it is open will seal it shut; pressing again will open it.

◆ If the rubber lip seal wears out, simply unscrew the stopper and replace it with a matching one (inset).

Time and heavy use take an inevitable toll on the hardware of a bathroom washbasin or a kitchen sink, but most problems can be readily corrected. For example, if mineral deposits clog a faucet aerator or spray head, the fitting can be cleared or replaced in mere minutes *(below and right)*. Even leaks from a dishwasher or disposer can often be repaired without professional help.

Leaks under a Fixture: Steady dripping or seepage beneath sinks and basins can be traced to any of several causes. The problem may

be cracked putty under a kitchen strainer *(opposite, bottom)*, or a worn hose *(pages 73 and 75)*, or a loose connection of one sort or another. Minor adjustments can sometimes solve the problem; for instance, leakage at the nuts that secure water supply tubes to the faucet may be stanched by tightening the nuts a little at a time with a wrench. In other cases, a new fitting is needed—perhaps an improved one: If, for example, a leak is due to the rusting of an old metal trap, remove it and put a modern plastic trap in its place

Fixing or Replacing Faucets: To repair a dripping spout or leaks around a faucet handle, purchase a kit of faucet replacement parts at a plumbing-supply store, specifying the make and model of the faucet. (Some common models are illustrated on pages 76 to 80.) If a faucet develops new leaks after the repair—or if you simply want a more modern fitting—remove the old faucet, clean the sink surface, and hook up the new one *(page 81)*. The job is not difficult, since the water supply tubes are generally quite accessible.

 TOOLS

Channel-joint pliers
Long-nose pliers
Basin wrench
Pipe wrench
Hammer and
  wooden dowel
Adjustable wrench

Small pliers
Screwdriver
Utility knife
Putty knife
Hex wrench
Seat wrench

 **MATERIALS**

Electrician's tape
Plumber's putty
Hose clamps
Penetrating oil
O-rings

Plumbing-sealant
  tape
Gaskets and
  washers
Silicone lubricant

Faucet repair kit
Faucet cartridge
Coupling nuts
Lock nuts
Supply tubes

SERVICING AERATORS AND SPRAY ATTACHMENTS

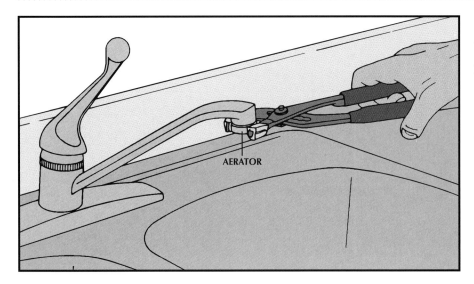

Fixing a clogged aerator.
◆ Wrap electrician's tape around each jaw of a pair of channel-joint pliers to prevent scarring of chrome parts, and unthread the aerator from the spout *(left)*.
◆ Soak the aerator in a mixture of water and white vinegar for a few minutes. Rinse the aerator and reattach it by hand, then tighten a quarter-turn with the pliers.
◆ Run water to check the aerator. If it is still clogged, remove it and install a new one, available at a plumbing-supply store.

AERATOR

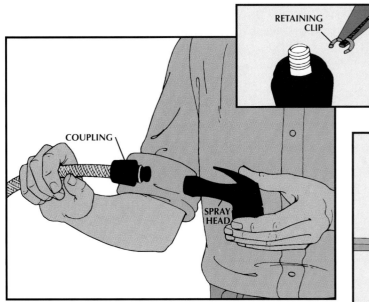

Servicing a blocked spray head.
◆ Unscrew the spray head from the hose coupling *(above).*
◆ Soak the spray head in water and white vinegar for a few minutes and rinse thoroughly. Reattach it.
◆ If the spray remains weak, replace the spray head and the coupling. To detach the old coupling, snap off its plastic ring and remove the retaining clip with a pair of long-nose pliers *(inset).*
◆ Secure the new coupling with a retaining clip. Attach the new spray head.

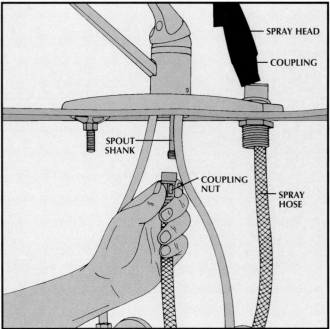

Replacing a leaking spray hose.
◆ With a basin wrench *(page 81),* loosen the coupling nut that attaches the spray hose to the spout shank.
◆ Unscrew the nut by hand and pull the hose free *(above).*
◆ If you plan to reuse the spray head, detach it from the old hose as in the repair at left.
◆ Connect the new hose to the spout shank.
◆ Pull the free end of the hose up through the hole in the faucet set. Attach a coupling and the spray head.

CURING SEEPAGE AROUND A KITCHEN STRAINER

1. Removing the strainer.
◆ Remove the sink trap *(page 42).* Then take out the pipe that leads from the trap to the strainer body.
◆ Loosen the lock nut at the base of the strainer body with a pipe wrench or, as shown at right, a hammer and wood dowel: Brace the dowel against one of the grooves of the lock nut and tap the dowel with a hammer.
◆ Push the strainer body out of the sink from underneath. If it appears worn, replace it.

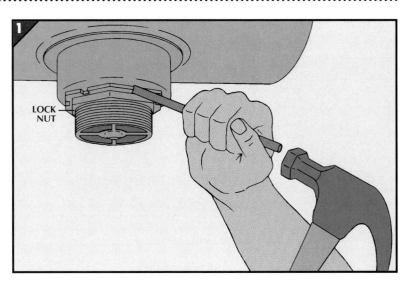

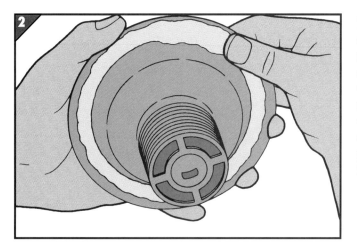

2. Sealing the opening.

◆ With a putty knife, clean old putty from the sink around the drain. If you are reusing the old strainer body, clear any old putty from it as well.

◆ Apply a $\frac{1}{8}$-inch bead of plumber's putty to the flange of the strainer body *(left)*.

◆ Insert the strainer in the opening and press down firmly so that the putty spreads evenly. Scrape away any excess putty around the opening with the putty knife, being careful not to scratch the surface.

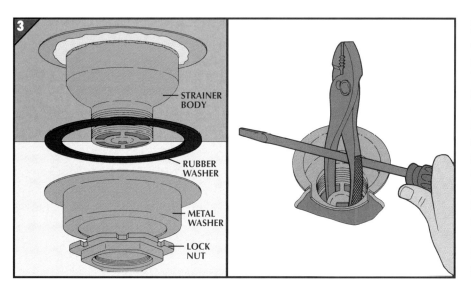

STRAINER BODY

RUBBER WASHER

METAL WASHER

LOCK NUT

3. Securing the strainer.

◆ Slip the rubber washer and the metal washer onto the strainer body. Tighten the lock nut by hand to hold the parts in place.

◆ Have a helper working from above insert the handles of pliers into the strainer crosspiece and slide a screwdriver between the handles *(near left)*.

◆ With a helper holding the screwdriver stationary so that the strainer body will not turn, tighten the lock nut with a pipe wrench or hammer and dowel *(page 73, Step 1)*. Do not overtighten the lock nut lest you distort the metal parts or crack the ceramic.

SOLVING GARBAGE DISPOSER LEAKS

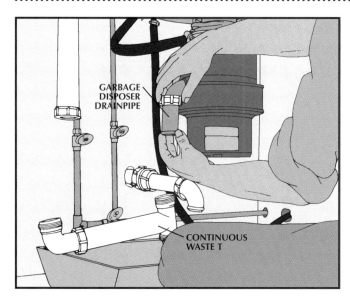

GARBAGE DISPOSER DRAINPIPE

CONTINUOUS WASTE T

Servicing the disposer drainpipe.

◆ Shut off power to the garbage disposer.

◆ Remove the sink trap *(page 42)*; in a double sink, also remove the continuous waste T, which carries wastewater from both sinks to the shared trap.

◆ Take out the screws in the fastener that attaches the disposer drainpipe to the disposer, then pull out the drainpipe *(left)*.

◆ Remove and replace the washer at the opening where the drainpipe connects to the disposer.

◆ Reassemble the pipes, replacing the slip nut washers as you work.

◆ Restore power to the disposer.

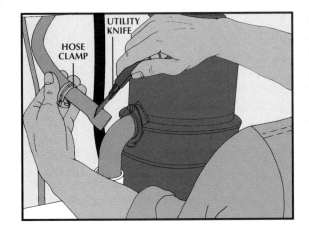

Repairing the dishwasher drain hose.

◆ If water is leaking from the drain hose between the dishwasher and disposer, shut off power to both appliances.

◆ Loosen the hose clamps and pull the hose off the disposer and the dishwasher. If only an end of the hose is damaged, cut away an inch or two with a utility knife *(left)*. Otherwise, buy a new hose.

◆ Slide new clamps onto the hose, push the ends of the hose onto the connections, and tighten the clamps.

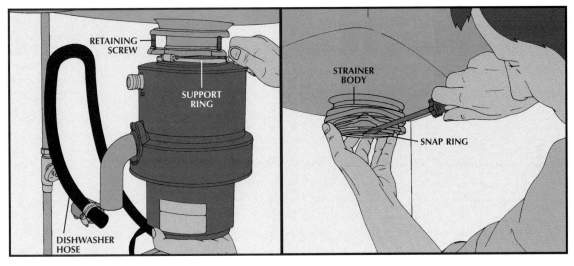

Fixing a leak at the disposer mounting assembly.

◆ Shut off power to the disposer and place a container under it.

◆ Remove the trap; for a double sink, also remove the continuous waste T, as when servicing the disposer drainpipe *(opposite page)*. Disconnect the dishwasher hose from the disposer.

◆ Supporting the disposer with one hand, push one of the lugs on the sup-port ring to turn the ring and unlock it from the mounting assembly *(above, left)*. If the ring is stubborn, use a screwdriver for extra leverage. Lower the disposer and set it aside.

◆ Remove the screws on the mounting assembly. Push up the flange at the base of the assembly and hold it as you pop out the snap ring with a screwdriver *(above, right)*. Take off the flange and its gasket, and push the strainer body up into the sink.

◆ Replace dried putty as explained on the opposite page, at top. Lower the strainer body into place and attach a new rubber gasket from below.

◆ Reinstall the mounting flange, the snap ring, and the retaining screws, then remount the garbage disposer. Attach the dishwasher hose and the drainpipes.

◆ Restore power to the disposer.

A REMEDY FOR A LEAKING WASHBASIN DRAIN

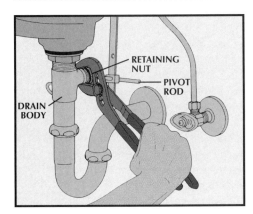

Tightening a pop-up plug connection.

The pop-up drain plug in a bathroom washbasin is operated by a pivot rod, typically linked through other components to a lift rod behind the faucet spout. If water leaks from the point where the pivot rod enters the drain body, tighten the rod's retaining nut with an adjustable wrench or channel-joint pliers *(left)*. If the leak continues, unscrew the retaining nut, slide the pivot rod out of the drain body, replace the washer or gasket under the nut, and reassemble.

DOUBLE-HANDLE FAUCETS

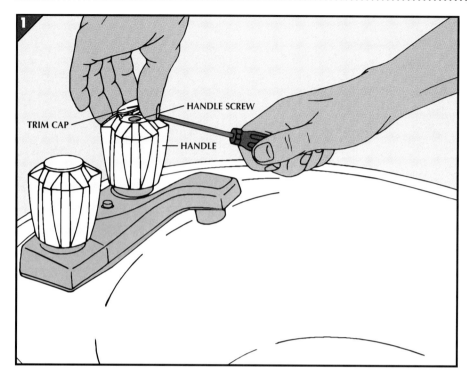

1. Opening up the faucet.

For a leaky handle, repair the faucet on that side. For a dripping spout, start by identifying the side to repair. Turn off one of the shutoff valves beneath the sink. If the leak stops, the problem is with that handle; if it persists, the other handle is at fault, or both are.

◆ With the water supply off, open the faucet, then close the handle when the water stops. Cover the drain to avoid losing small parts.

◆ Carefully pry off the trim cap with a knife or small screwdriver *(left),* and remove the screw below the cap.

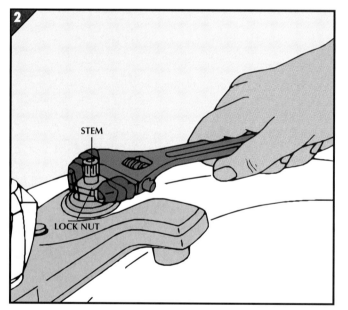

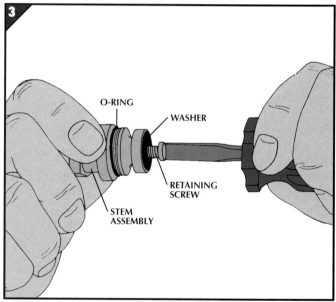

2. Getting at the stem.

◆ Lift off the faucet handle and the sleeve inside it, if there is one. If the handle will not budge, apply penetrating oil and wait an hour before trying again. Never strike the handle or sleeve with a hammer; this could damage the soft brass stem.

◆ Open the faucet stem one half-turn with tape-wrapped pliers, then use an adjustable wrench to unscrew the lock nut that secures the stem to the faucet body *(above).*

3. Repairing the stem assembly.

◆ Grasp the stem with taped pliers and lift it out of the faucet body, much as for a tub faucet stem *(page 61).*

◆ To correct a drip at the spout, replace the washer at the base of the stem. It may be held in place with a retaining screw *(above)* or a spring, or it may be shaped to fit around the base.

◆ To stop a leak around the handle, pinch off the O-ring. Lubricate a new O-ring with soapy water or silicone lubricant and roll it onto the stem until it is firmly seated.

◆ Reassemble the faucet and open the shutoff valve. If the leak continues, replace the seat *(Step 4).*

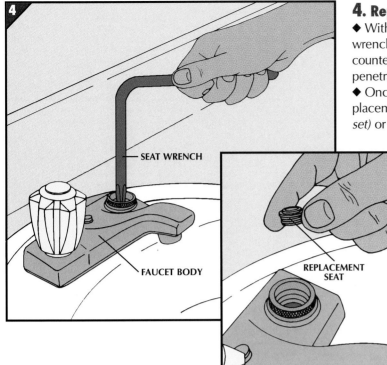

SEAT WRENCH

FAUCET BODY

REPLACEMENT SEAT

4. Replacing the seat.

◆ With a hex wrench or a special faucet seat wrench *(left)*, unscrew the seat by turning it counterclockwise. If the seat will not turn, apply penetrating oil, wait overnight, and try again.

◆ Once the old seat is free, fit an identical replacement seat into the faucet body by hand *(inset)* or with a pair of long-nose pliers.

◆ Screw it in tightly with the hex wrench or seat wrench.

◆ Reassemble the faucet and turn on the water supply.

REPAIRS FOR SINGLE-LEVER CARTRIDGE FAUCETS

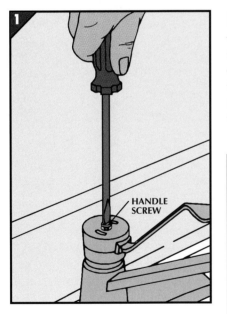

HANDLE SCREW

1. Gaining access to the handle.

◆ Turn off the water to the faucet, lift the handle several times to drain the faucet, and close the drain to prevent loss of small parts.

◆ Carefully pry off the trim cap with a small screwdriver or knife, exposing the handle screw.

◆ Remove the handle screw *(left)*.

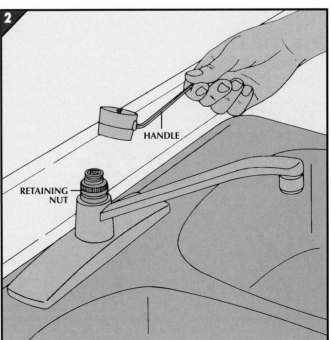

HANDLE

RETAINING NUT

2. Removing the handle.

The faucet handle attaches to the lip of the retaining nut below it much like a bottle opener to a cap. Tilt the handle lever up sharply to unhook it from the nut, then lift it free.

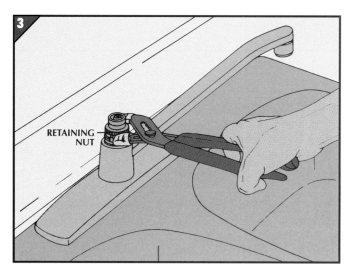

3. Removing the retaining nut.

Unscrew the retaining nut with taped channel-joint pliers and lift it off the faucet body.

4. Repairing a drip or leak.

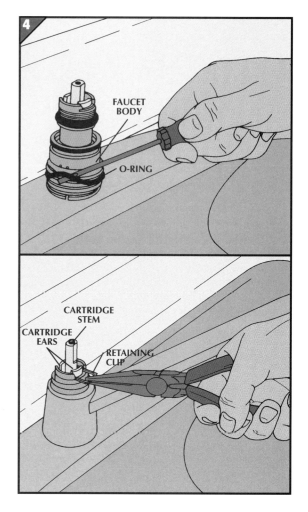

◆ For a leak at the spout collar, lift off the spout, then pry the O-rings off the faucet body *(right, top)*. Lubricate new O-rings with soapy water or silicone lubricant and roll them into place. Reattach the spout, retaining nut, and handle, hooking the inside edge of the handle on the lip of the retaining nut.

◆ For a dripping spout or a leak at the handle base, leave the spout in place. Remove the cartridge retaining clip *(right, bottom)*. Note the cartridge ears' orientation, then lift out the cartridge as for the tub faucet on page 63. If the cartridge is worn, replace it; otherwise replace only the O-rings. Reinsert the cartridge, aligning the ears as before, and attach the retaining nut and handle. If the hot and cold water are reversed, remove the handle and nut and rotate the cartridge stem a half-turn. Reassemble the faucet and open the shutoff valves.

SERVICING SINGLE-LEVER BALL FAUCETS

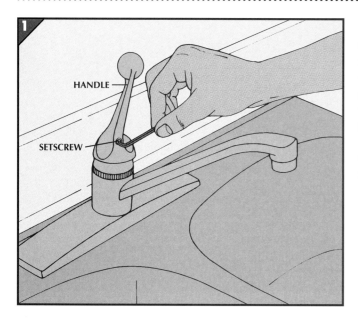

1. Removing the handle.

◆ Turn off the water supply and open the faucet. Then cover the drain.

◆ With a hex wrench, loosen the setscrew that secures the handle to the faucet body *(left)*. Because the setscrew is small and easily lost, leave it in the handle.

◆ Lift off the handle to expose the cap below.

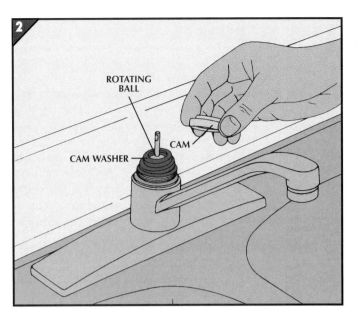

2. Removing the cam assembly.

◆ Remove the cap as explained on page 62 for a tub faucet that is of the same type.

◆ Lift off the plastic cam *(left)*, exposing the cam washer and rotating ball.

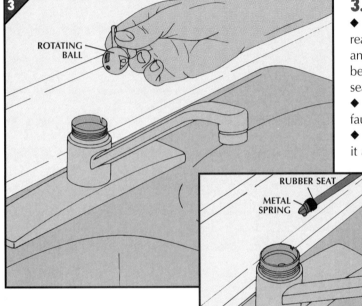

3. Replacing the seats, springs, and ball.

◆ Lift the rotating ball from the faucet body *(left)*, then reach into the body with long-nose pliers or a screwdriver and remove the components inside: either two sets of rubber seats and metal springs *(inset)* or two sets of ceramic seals and O-rings.

◆ Replace the seats and springs from a repair kit for the faucet make and model.

◆ If the ball-and-cam assembly appears damaged, replace it as well.

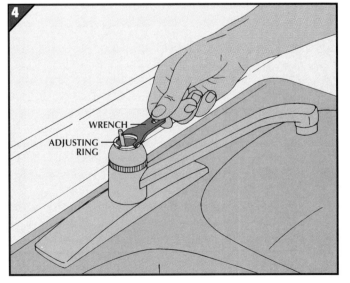

4. Tightening the adjusting ring.

◆ Replace the cap over the rotating ball, tightening it with channel-joint pliers taped to avoid scarring.

◆ With a special wrench included in the repair kit, tighten the adjusting ring clockwise *(left)*. The ball should move easily without the handle attached.

◆ Reassemble the handle and turn on the water supply.

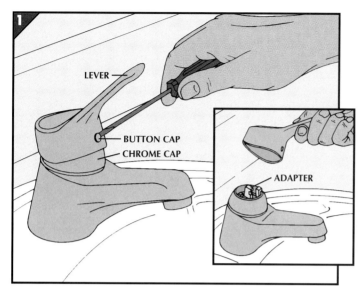

1. Removing the lever.

◆ Turn off the water supply. Drain the faucet by lifting the lever to its highest position. Close the drain to prevent loss of parts.

◆ Pry off the button cap at the base of the lever with a knife or small screwdriver *(left),* and remove the handle screw underneath the cap. (On some models the screw is under the lever body and there is no cap.)

◆ Lift off the lever *(inset),* exposing the chrome cap below. Pry the cap off its plastic adapter or, on some faucets, unscrew it from the faucet body.

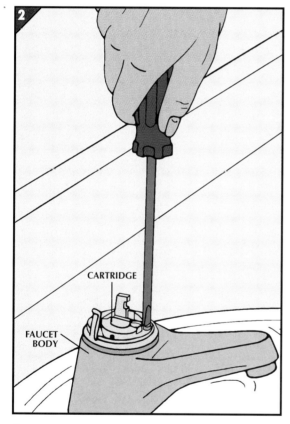

2. Freeing the cartridge.

◆ Loosen the two or three brass screws that hold the cartridge to the faucet body *(above).*

◆ Lift out the cartridge.

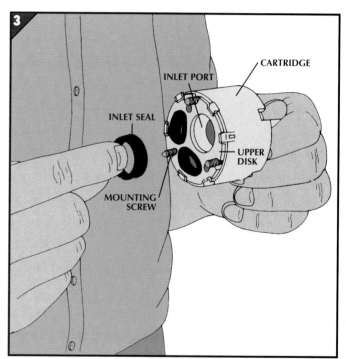

3. Servicing the cartridge.

◆ Leaks are sometimes caused by a piece of dirt caught in the faucet mechanism. Clean the inlet ports of the upper disk and the surface of the bottom disk, which is permanently mounted inside the faucet body.

◆ If the upper disk is cracked or pitted, buy a replacement cartridge for the same make and model of faucet.

◆ Insert new inlet seals in the upper disk *(above).*

◆ Position the cartridge so that the three ports on the bottom align with those of the faucet body.

◆ Secure the cartridge in place with screws and reattach the chrome cap and lever.

◆ Open the shutoff valves.

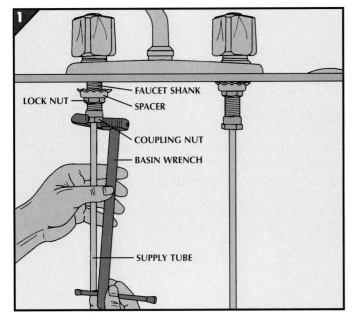

1. Removing the old faucet set.

◆ Shut off the sink water supply, open the faucet to empty it, and cover the drain.

◆ Working under the sink, use a basin wrench to loosen the lock nut at the top of a faucet shank and the coupling nut between the shank and the supply tube *(left)*. If a nut is stuck, apply penetrating oil and wait at least an hour before trying again. Detach the nuts and supply tube from the shank and remove the spacer, if any.

◆ Repeat the procedure for the other faucet shank.

◆ Lift the faucet set out of the sink.

◆ Scrape old putty off the sink surface and remove mineral deposits with white vinegar.

◆ Unless you plan to reuse the supply tubes, loosen the nuts securing them to the shutoff valves and replace them with new, flexible supply tubes.

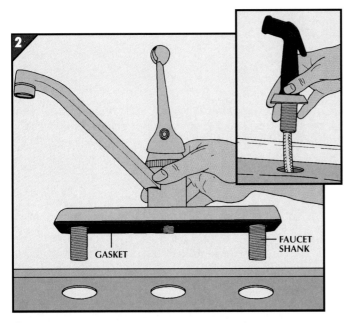

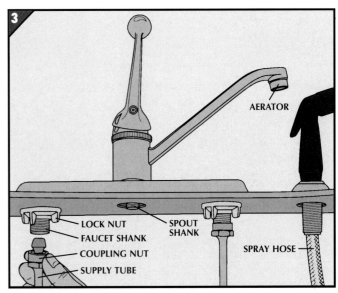

2. Seating the new faucet.

◆ Buy a faucet with the same number and spacing of shanks as the old one, so that it fits the holes in the sink. For a kitchen sink, also buy a spray attachment if there is an extra hole for one; a hole not already in use may be concealed by a metal plate.

◆ Seat the faucet's rubber gasket snugly on the bottom of the faucet. Then set the faucet in place *(above)*.

◆ Wrap each faucet shank with plumbing-sealant tape and secure it to the sink with a lock nut. Tighten a plastic nut by hand, a metal nut with a basin wrench.

◆ If you are installing a spray attachment, feed the hose through the seating hole *(inset)*.

3. Connecting the faucet.

◆ Attach the supply tubes to the faucet shanks with coupling nuts *(above)*, tightening each a half-turn with a basin wrench.

◆ Attach the spray hose, if any, to the spout shank with a coupling nut *(page 73)*.

◆ Remove the aerator *(page 72)*. Uncover the drain and restore the water supply.

◆ Flush the faucet by turning it first to hot and then to cold. As the water runs, check under the sink for leaks. If any nut leaks, tighten it a little at a time.

◆ Turn off the faucet and reinstall the aerator.

Toilet Repair and Installation

Toilets, the most heavily used plumbing fixtures in the home, routinely require small repairs and adjustments—most of which can be performed in minutes. Replacing a toilet is only somewhat more demanding. A homeowner can usually complete the task in an afternoon *(pages 89-91)*.

Common Problems: When the working parts inside a toilet tank wear out or corrode, the likely result is either a continuously running toilet or an inadequate flush. A weak flush can also mean flushing holes are blocked. Diagnose and correct the specific cause as shown here and on pages 84 to 87.

To identify a costly but hidden leak from tank to bowl, perform the test on page 51 and replace the flapper ball if necessary *(opposite, bottom)*. Visible leaks from the tank usually result from loose bolts or worn washers and gaskets *(page 84)*. Finally, clear a clogged toilet with a fold-out plunger or closet auger, as described on page 41.

Pressure-flush toilets are more likely to need the services of a professional. You can sometimes correct a weak flush yourself, though, as shown on page 88.

Avoiding Complications: Whether repairing or installing a toilet, place the tank lid in a safe place. Also be careful in tightening bolts. Porcelain cracks easily, and when it does the toilet usually must be replaced.

Some toilets have plastic nuts. These are made to be turned by hand and could be cracked or crushed by a wrench. Use a wrench only when an old plastic nut cannot be removed by hand and you do not plan to reinstall the nut.

 TOOLS

Adjustable wrench	Locking-grip pliers
Long-nose pliers	Diagonal pliers
Screwdriver	Putty knife
Spud wrench	Level

 **MATERIALS**

Sponge	Penetrating oil	Pressure-flush fittings
Flapper ball	Special-purpose gaskets	Closet bolts and nuts
Lift chain	Spud washer	Closet-bolt caps
Plumbing-sealant tape	Float-cup ball cock	Sheet metal (for shims)
Washers, bolts, nuts	Float ball	Caulk

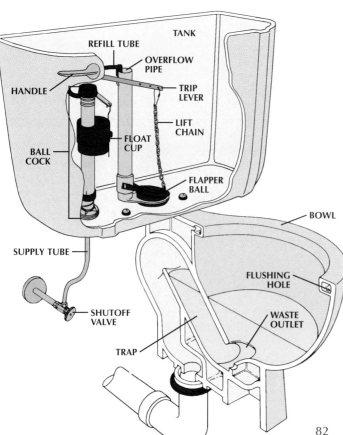

How a standard toilet works.

In a typical modern toilet *(left)*, a flapper ball seals the hole between the tank and the bowl except during a flush cycle. When the handle is pressed, it operates a trip lever and chain that raise the ball. Water rushes into the bowl through flushing holes inside the rim and then out the waste outlet. As the water level falls in the tank, a float cup buoyed by the water drops as well. The float cup is part of a mechanism called a ball cock, and the cup's descent triggers the ball cock to let in water from the toilet's supply tube. Most of the water flows into the tank, but some is directed through a refill tube to replenish the bowl and the toilet's internal trap. After the flush, the flapper ball settles back into place, resealing the tank. The float cup then rises with the incoming water, shutting off the ball cock when the water level is back to normal. If some maladjustment causes the tank water to rise too high, water spills through the overflow pipe into the bowl.

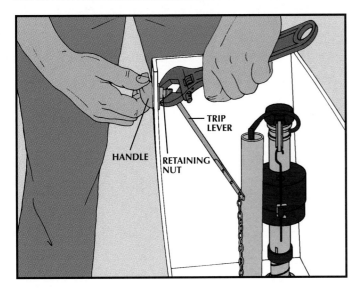

Adjusting the handle.

When a handle must be held down to complete a flush, the cause may be a loose connection to the trip lever. Grasp a loose handle and tighten the retaining nut inside the tank by turning it counterclockwise (the opposite direction from most nuts), with your hand for a plastic nut and with an adjustable wrench for a metal one *(left).*

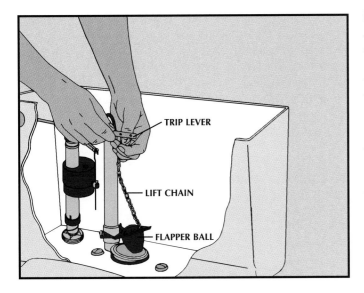

Servicing the lift chain.

Another cause for a handle that must be held down is a lift chain that is too long. To shorten it, unhook the chain from the trip lever and adjust it to allow about $\frac{1}{2}$ inch of slack: Hook the chain through another hole in the lever *(left)* or use long-nose pliers to cut and remove links.

A slow flush may indicate a short or broken chain. Replace it; pin or paper clip repairs cause corrosion. Some chains have a small float attached that regulates the amount of water in each flush to match a level marked inside the tank. If the volume of the flush is too low or too high, adjust it by sliding the float along the chain.

Replacing the flapper ball.

Stop tank-to-bowl leaks with a new flapper ball:
◆ Turn off the water, flush the toilet, and sponge out the tank.
◆ Unhook the ball's lift chain and take the refill tube out of the overflow pipe. Remove the flapper ball by sliding the collar up and off the pipe.
◆ To ensure a good seal, wipe a plastic valve seat with a soft cloth; gently scour a metal one with fine steel wool or a plastic cleansing pad.
◆ Slide a new flapper ball and collar down the overflow pipe. Position the ball and screw the collar in place.
◆ Attach the lift chain from the ball to the trip lever; reseat the refill tube in the overflow pipe.
◆ Turn the water back on.

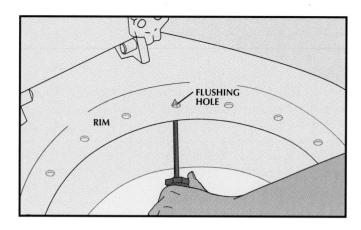

Clearing the flushing holes.

If water flows unevenly down the walls of the bowl during a flush, mineral deposits may have clogged some of the flushing holes under the rim. Insert the blade of a small screwdriver or pocket knife into the flushing holes and scrape them clean. If there is a small hole opposite the waste outlet at the base of the bowl, turn off the water, flush the toilet, and scrape inside that hole as well; then turn the water back on.

STOPPING LEAKS

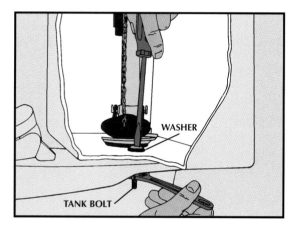

Sealing leaks under the tank.

Condensation can cause water drips on the floor under the tank. More often, the cause is a leak from the tank.

◆ To fix a leaking tank bolt, secure it inside the tank with a screwdriver and tighten the nut with an adjustable wrench *(left)*. If leaks persist, replace the bolt washers.

◆ Tighten a loose lock nut at the base of the ball cock: Turn a plastic nut with your hand, a metal nut a quarter-turn at a time with an adjustable wrench. If a finger-tight plastic nut still leaks, remove it and wrap plumbing-sealant tape on the exposed threads, then reattach the nut.

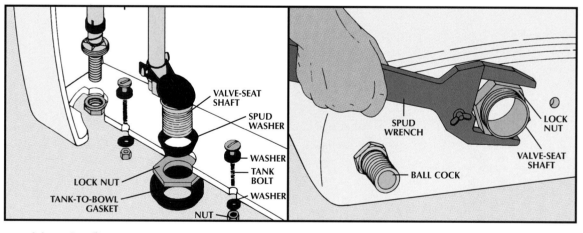

Servicing the flush-valve connection.

Pooled water near where the bowl meets the tank may be a sign of a poor tank-to-bowl connection *(above, left)*. To fix it, you must remove the tank.

◆ Shut off the water, flush the toilet, and sponge out the tank.

◆ With an adjustable wrench, loosen the coupling nut at the top of the supply tube and detach the tube from the tank.

◆ Remove the tank bolts, reversing the tightening procedure shown above.

◆ Lift the tank straight up and off; set it aside. Pry

or scrape the gasket from the tank and bowl.

◆ Turn the lock nut with a spud wrench *(above, right)*. If necessary, apply penetrating oil.

◆ Pull the valve-seat shaft into the tank. Replace the old spud washer, then push the shaft back into place, securing it with the lock nut. Push a new tank-to-bowl gasket up over the nut.

◆ Ease the tank onto the bowl and attach it with new tank bolts, nuts, and washers. Reconnect the supply tube.

◆ Turn on the water slowly. Tighten any leaking nuts a quarter-turn.

REPLACING THE FILL MECHANISM

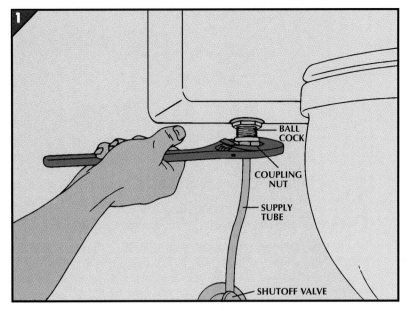

1. Detaching the supply tube.

You can often identify a worn-out ball cock by a whistling sound or tank vibration after a flush. Possible other symptoms include a low tank water level or, conversely, nonstop filling. Replace the ball cock as follows:

◆ Set a container on the floor beneath the tank to catch water runoff.

◆ Turn off the water supply. Flush the toilet and sponge out the tank.

◆ Disconnect the supply tube from the base of the ball cock by loosening the coupling nut, turning a plastic nut by hand and a metal one with an adjustable wrench *(left)*.

◆ Free the top of the ball cock by unclipping the refill tube from the overflow pipe. For an old-style ball cock *(page 87)*, also take off the float rod.

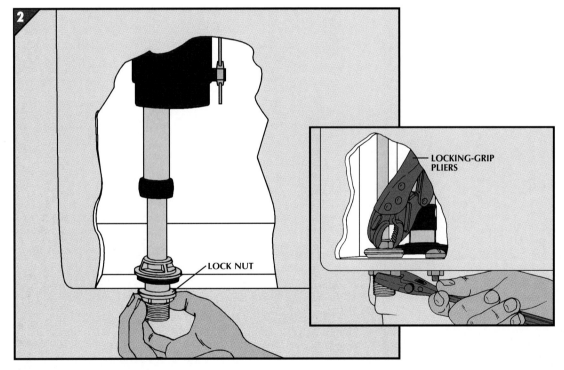

2. Removing the old ball cock.

◆ Working under the tank, unscrew a plastic lock nut at the base of the ball cock by hand *(above)*. Discard the nut.

◆ For a plastic nut that will not turn, or for a metal lock nut, attach locking-grip pliers to the lowest part of the ball cock inside the tank and wedge them against the tank wall to prevent the ball cock from turning; then remove the lock nut with an adjustable wrench *(inset)*. Apply penetrating oil to a metal nut that does not turn.

◆ Pull the ball cock up and out.

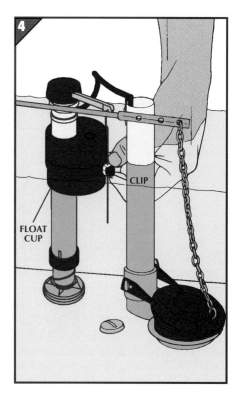

3. Installing a ball cock.

◆ Insert a new ball cock into the tank hole *(above)*, making sure it stands straight without touching the sides. Clip the refill tube to the overflow pipe.

◆ Holding the ball cock with one hand, slip a washer and plastic lock nut over the base of the ball cock under the tank. Tighten the lock nut by hand.

◆ Reconnect the supply tube and slowly turn on the water.

◆ If a leak occurs, do not tighten plastic nuts with a wrench; instead, disassemble the connections and wrap plumbing-sealant tape around the exposed threads before reattaching.

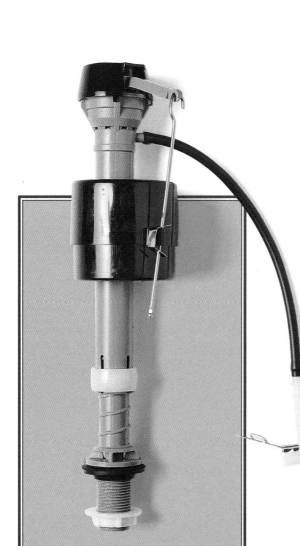

A SAFER BALL COCK

Household water pressure can drop suddenly—for example, if a local main breaks. Toilet tank water, including any chemical cleansers, may then be siphoned backward into the neighborhood water supply. To prevent that, many ball cocks are sold with an antisiphon device; in the model shown, an air intake stops siphon action. Most local codes require such features, but ball cocks without them are available. Make sure the packaging for yours specifies antisiphon protection.

4. Adjusting the float cup.

◆ Flush the toilet, let the tank refill, and check the water level; it should be $\frac{1}{2}$ to 1 inch below the top of the overflow pipe and should be just below the handle.

◆ Adjust the level by pinching the clip at the side of the float cup and sliding it $\frac{1}{2}$ inch at a time—up to raise the water level or down to lower it.

REPAIRS FOR A FLOAT-BALL ASSEMBLY

Although float-cup ball cocks like the one at left are common, some older toilets have float-ball assemblies like the one shown below. In that arrangement, the flush mechanism and the water-supply connection are the same as for any standard toilet, but the fill mechanism is quite different.

The water level in such a toilet is regulated by a float ball and rod, rather than a float cup. When the tank empties during flushing, the ball drops with the falling water level. This lowers the rod, which in turn opens the ball cock, allowing water into the tank. The float ball is then buoyed upward as the tank refills. At the correct water level, the float rod turns off the ball cock.

If the water level is too high in a toilet that contains this mechanism, first check the float ball. A cracked or worn ball can partially fill with water, preventing it from rising enough to shut off the ball cock. Replace a float ball by grasping the float rod firmly with a pair of pliers and unscrewing the ball. Attach a new ball in the same way.

If the float ball is in good order, try adjusting the float rod slightly downward— about $\frac{1}{2}$ inch. Some plastic versions have a dial that can be turned to correct the height of the arm. Bend a metal float rod; if doing so proves difficult inside the tank, unscrew the rod and bend it over a rounded surface.

You can also replace a float-ball assembly and ball cock with a float-cup ball cock; it is less likely to need repairs *(pages 85-86)*.

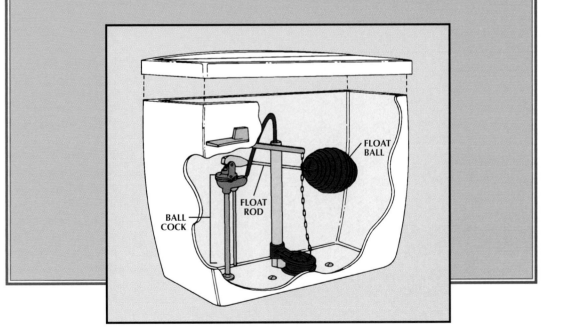

FLOAT BALL

FLOAT ROD

BALL COCK

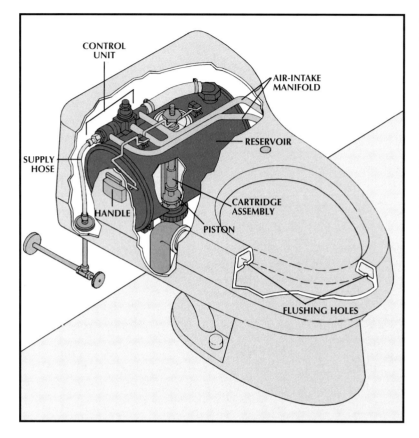

Anatomy of a pressure-flush toilet.

Molded one-piece toilets like the one at left often include a pressure-flush mechanism. Between flushes, a reservoir holds air and water at a pressure of about 25 pounds per square inch. A cartridge assembly seals an outlet at the reservoir's base. Pressing the handle lifts the cartridge assembly, raising the piston and flushing the toilet with water and air discharged through holes under the bowl rim. The piston then reseals the reservoir. Triggered by reduced reservoir pressure, a regulator located inside a control unit takes in water from the supply hose. The water and air from an intake manifold proceeds through an inlet hose into the reservoir, fully recharging it in less than 90 seconds.

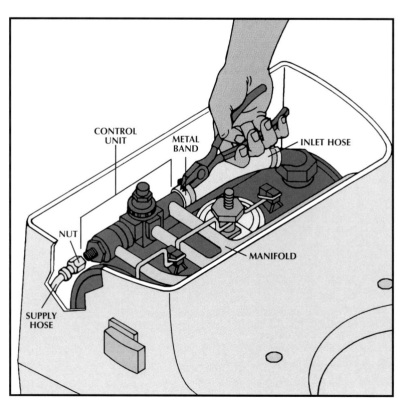

Restoring a weak flush.

When a pressure-flush system fails, water often continues to flow, but very weakly. Replacing the control unit may solve the problem:

◆ Close the toilet shutoff valve.

◆ Flush the toilet to release any remaining pressure in the reservoir.

◆ Loosen the nut connecting the supply hose to the control unit. With diagonal cutting pliers, sever the metal band that secures the reservoir inlet hose to the control unit *(left)*, then pull the control unit out of the manifold openings.

◆ Install a new control unit on the manifold.

◆ Reconnect the supply hose to the control unit and tighten the nut.

◆ Slip a hose clamp over the inlet hose, then slide the hose onto the control unit and tighten the clamp.

◆ Slowly open the shutoff valve and test the toilet. If the flush is not restored, consult a plumber with expertise in pressure-flush systems.

PUTTING IN A MODERN TOILET

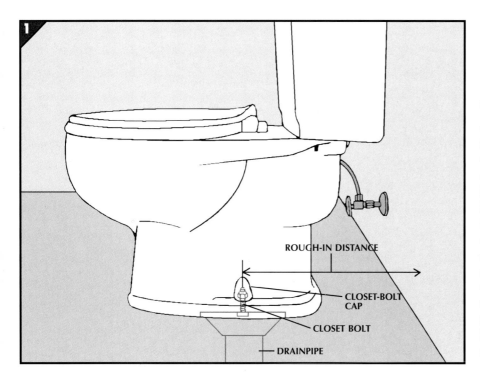

1. Finding the rough-in distance.
◆ Locate the closet bolts that secure the old toilet bowl to the floor.
◆ Measure from the center of the closet-bolt caps to the wall behind the bowl *(left);* if there are four closet bolts, measure from the rear pair. This is the toilet's rough-in distance—typically 10, 12, or 14 inches—which indicates the location of the concealed drainpipe in relation to the wall.
◆ Buy a new toilet with a rough-in distance that is either the same as or less than this measurement. It must never be longer or there will not be enough space for the new fixture.

2. Taking off the tank.
◆ Close the shutoff valve, flush the toilet, and bail out and sponge water from the tank and bowl.
◆ Following the procedures on pages 84 and 85, remove the tank bolts and detach the supply tube. Disconnect the other end of the supply tube from the shutoff valve and discard the tube.
◆ Lift the tank off the bowl *(right).*

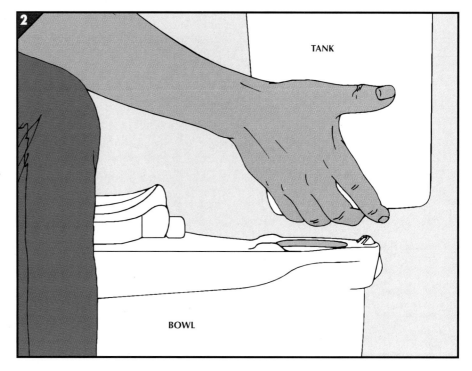

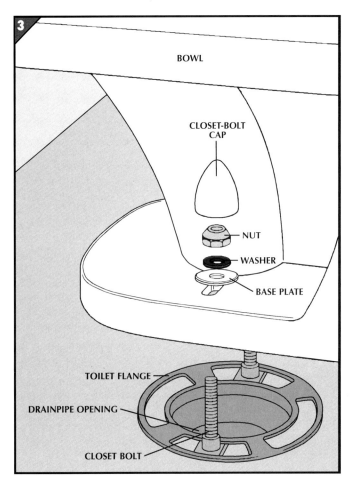

3. Removing the bowl.

◆ Unscrew or pry off the closet-bolt caps, and remove the nuts and washers below. For nuts that resist turning, apply penetrating oil and wait 15 minutes before trying again.

◆ To break the seal between the toilet bowl and the floor, grasp the bowl and twist or rock it back and forth. Then lift the bowl straight up off the bolts and set it aside.

◆ Remove the bolts from the flange. Stuff a rag into the drainpipe opening to keep out sewer gas and prevent debris from falling in.

◆ With a putty knife, scrape away the old wax gasket from the toilet flange.

◆ Examine the flange to make sure it is not broken. If it is, a new flange may be necessary.

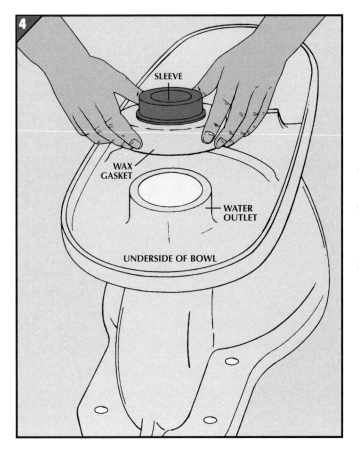

4. Installing the wax gasket.

With the new bowl upside down, place a wax gasket around the water outlet, using a gasket with a sleeve *(left)* for a toilet flange that is recessed slightly below floor level. Install the gasket with the sleeve facing away from the water outlet.

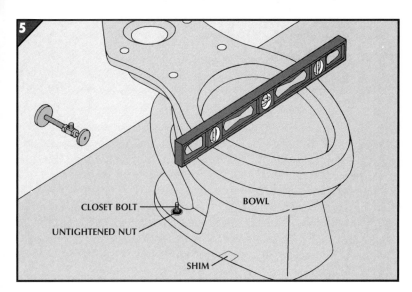

5. Seating the new bowl.

◆ Remove the rag from the drain. Insert closet bolts in the flange, placing each the same distance from the wall.

◆ Turn the bowl upright and position it above the flange, aligning the holes at the base of the bowl with the bolts.

◆ Lower the bowl into place and press down to ensure a good seal between bowl and drainpipe. Do not break the seal by lifting the bowl again.

◆ Use a level to be sure the bowl is not tilted. If necessary, slide thin sheet-metal shims under the bowl base *(left)*.

◆ For plastic bolt caps, slip a cap base, washer, and nut onto each bolt; if the caps are porcelain, place only a washer and nut on the bolt. Screw the nuts on but do not tighten them yet.

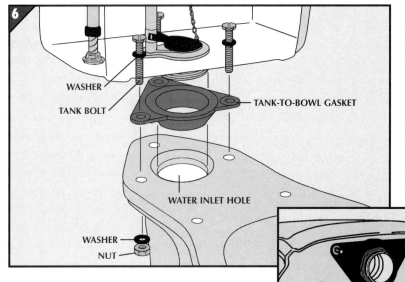

6. Adding the tank.

◆ Lower the new tank onto the bowl, aligning the center hole in the tank-to-bowl gasket with the water inlet hole and the corners of the gasket with the bolt openings on each side of the hole *(left)*. Modern toilets often have three tank bolts, rather than two.

◆ Insert tank bolts with rubber washers through the base of the tank. Fasten the bolts with nuts and washers where they emerge under the bowl.

◆ Adjust the toilet so the tank is parallel to the wall; you can twist it slightly without breaking the wax seal. Check that the bowl is level and does not rock; shim as needed.

◆ Tighten the nuts on the closet bolts. Snap on plastic caps, or press on porcelain caps after filling them with plumber's putty.

◆ Seal the bowl to the floor with caulk.

◆ Attach the seat and cover.

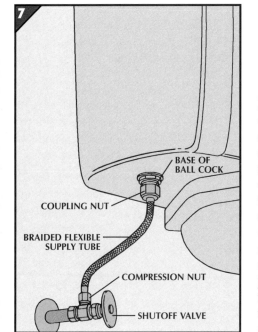

7. Connecting the water supply.

◆ Screw the compression nut at one end of a flexible braided supply tube to the shutoff outlet; fasten the other end of the tube to the ball cock with a coupling nut.

◆ Make plastic nut connections finger tight; with those for metal nuts, add a quarter-turn with an adjustable wrench.

◆ Open the shutoff valve slowly. To correct a leak at a metal nut, tighten the nut a fraction of a turn at a time with an adjustable wrench. For a leak at a finger-tight plastic nut, disassemble the connection and wrap plumbing-sealant tape around any exposed threads.

The Ins and Outs of Water Heaters

Water heaters are an essential part of any plumbing system. To keep a water heater in good order, perform the maintenance tasks on pages 95 and 96. If something goes wrong despite your efforts at prevention, you can often solve the problem yourself. Consult the troubleshooting chart below, then perform the repair as shown in the pages that follow.

Two Kinds of Heaters: In a gas-fired water heater, a burner is lit—either electrically or by means of a small gas flame called a pilot light—whenever there is a significant drop in the water temperature. Electric ignition systems rarely need service, but strong drafts sometimes extinguish the pilot light. Instructions for relighting it appear on pages 94 and 95.

An electric water heater substitutes two heating elements inside the tank. In areas with hard water, these elements can become caked with mineral deposits and may need periodic replacement *(page 100)*.

Putting in a New Heater: Water heaters typically last 5 to 15 years. Replacing a gas water heater is best left to a professional, but you can install a new electric water heater in a few steps *(pages 100-103)*.

 *Turn off power at the service panel before working on an electric heater.* **CAUTION**

 TOOLS

Screwdriver
Pipe wrench
Adjustable wrench
Continuity tester
Voltage tester
Pipe cutter
Propane torch
Flameproof pad

 **MATERIALS**

Pipe insulation
Plumbing-sealant tape
Pipe fittings
Electric heating element
Solder, flux
Copper pipe ($\frac{3}{4}$")
Wire caps

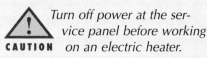 **SAFETY TIPS**

Gloves and safety goggles provide protection when soldering copper pipe.

Troubleshooting Guide

PROBLEM	REMEDY
No hot water.	Gas heater only: If the pilot is out, relight it *(pages 94-95)*; when there are repeated outages, check for floor drafts. Otherwise, test the thermocouple *(pages 96-97)*. Have the gas company check the control unit and burner. Electric heater only: Test heating elements; replace faulty ones *(page 100)*. If the elements are good, have a technician check the thermostats and high-temperature cutoff.
Not enough hot water.	Stagger household use of hot water, or install a larger heater *(pages 100-103)*. Drain the tank to remove sediment *(page 95)*. Insulate hot-water pipes *(page 96)*. Electric heater only: Test and replace heating elements *(page 100)*. If the elements are good, have a technician test the thermostats and cutoff.
Water too hot.	Lower the temperature setting *(pages 95, 103)*.
Water drips from discharge pipe.	Lower the temperature setting *(pages 95, 103)*. Service and if necessary replace the relief valve *(pages 97-98)*. Electric heater only: Have a technician test the thermostat and cutoff.
Drain valve leaks.	Service the drain valve and replace if necessary *(pages 98-99)*.
Water heater makes rumbling, popping, or crackling sounds.	Drain and refill the tank regularly to remove sediment *(page 95)*. Service the relief valve *(pages 97-98)*. Electric heater only: Replace heating elements coated with mineral scale *(page 100)*. Gas heater only: Have the gas company check the burner and gas pressure.

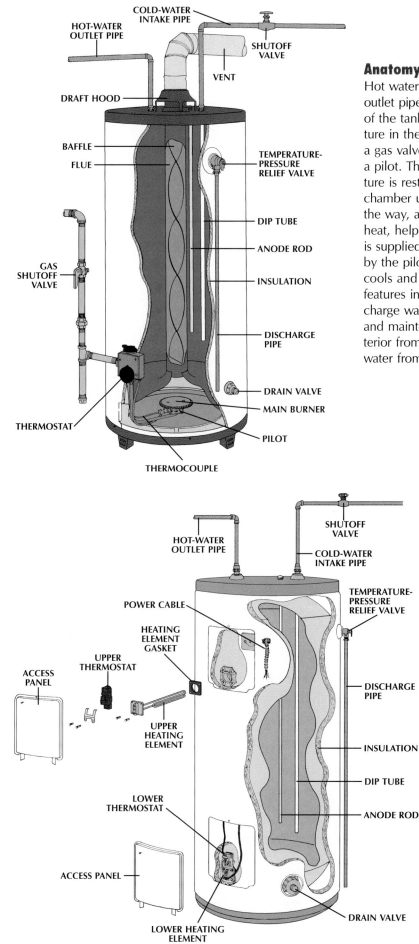

HOT-WATER
OUTLET PIPE

COLD-WATER
INTAKE PIPE

SHUTOFF
VALVE

VENT

DRAFT HOOD

BAFFLE

FLUE

TEMPERATURE-
PRESSURE
RELIEF VALVE

DIP TUBE

ANODE ROD

GAS
SHUTOFF
VALVE

INSULATION

DISCHARGE
PIPE

DRAIN VALVE

MAIN BURNER

THERMOSTAT

PILOT

THERMOCOUPLE

Anatomy of a gas water heater.

Hot water flowing out of the tank through the hot-water outlet pipe is replaced by cold water entering the bottom of the tank through a dip tube. When the water temperature in the tank drops below the level set at a thermostat, a gas valve opens to supply a burner, which is ignited by a pilot. The burner heats the tank until the water temperature is restored. Meanwhile, hot air passes from the burner chamber up through a flue, a draft hood, and a vent; on the way, a baffle inside the flue captures some of the heat, helping to warm the water further. The burner pilot is supplied with gas by a thermocouple that is kept warm by the pilot itself. If the pilot blows out, the thermocouple cools and shuts off the flow of gas. Other water-heater features include: a temperature-pressure relief valve to discharge water if the tank overheats; a drain valve for repairs and maintenance; an anode rod that protects the tank interior from corrosion; and insulation to keep the stored water from losing heat too quickly.

SHUTOFF
VALVE

HOT-WATER
OUTLET PIPE

COLD-WATER
INTAKE PIPE

POWER CABLE

TEMPERATURE-
PRESSURE
RELIEF VALVE

HEATING
ELEMENT
GASKET

UPPER
THERMOSTAT

ACCESS
PANEL

DISCHARGE
PIPE

UPPER
HEATING
ELEMENT

INSULATION

LOWER
THERMOSTAT

DIP TUBE

ANODE ROD

ACCESS PANEL

DRAIN VALVE

LOWER HEATING
ELEMENT

An electric water heater design.

As in a gas water heater *(above)*, hot water leaves an electric water heater tank from the top and is replaced by cold water that enters the bottom of the tank through a dip tube. A thermostat there senses the drop in temperature and turns on the lower element. When a great deal of hot water is used, cool water rises in the tank to the level of a second thermostat, which diverts power from the lower heating element to another element, shown here withdrawn from the tank. Inside the upper thermostat, a high-temperature cutoff preset at 190°F turns off the power if the water overheats. Like gas-fired models, an electric water heater also has an anode rod, drain valve, and temperature-pressure relief valve.

LIGHTING A GAS BURNER

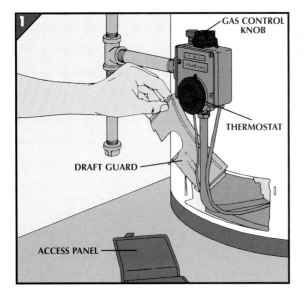

1. Gaining access to the pilot.
◆ Remove the burner access panel and draft guard. Some are detached by lifting them *(left)*; others slide sideways.
◆ Adjust the thermostat to its lowest setting and turn the gas control knob off.

If you smell gas, wait at least 5 minutes for the odor to clear. If it lingers, close the gas shutoff valve to the water heater, ventilate the room, and call the gas company. Otherwise, proceed to Step 2.

2. Setting the control knob.
Turn the control knob to the position labeled PILOT *(right)*.

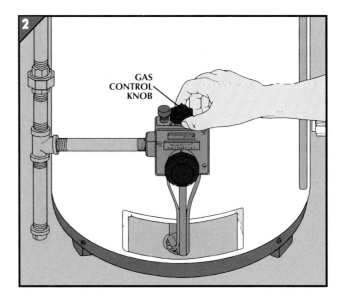

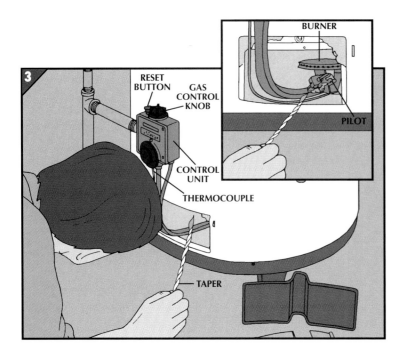

3. Lighting the pilot.
◆ Light a long fireplace match or a taper made from tightly rolled paper. Turn your head to one side and insert the lighted match or taper into the opening to check that there is no remaining gas there.
◆ Looking under the heater *(left)*, position the flame next to the pilot *(inset)*, depress the reset button, and hold it down; when there is no reset button, hold down the gas control knob. If the pilot fails to light within 5 seconds, close the gas shutoff valve and call the gas company.
◆ After the pilot lights, hold the reset button or gas control knob for 1 minute to heat the thermocouple.

If the pilot goes out when you release the button or control knob, turn off the gas control knob. Tighten the nut that connects the thermocouple to the control unit finger tight, then add a quarter-turn with a wrench. Should that repair fail, replace the thermocouple *(pages 96-97)*.

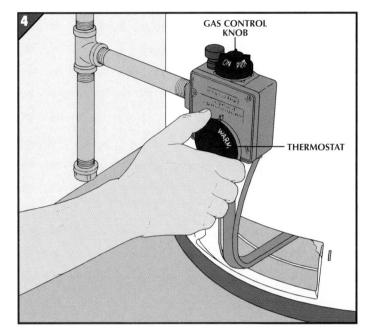

4. Setting the temperature.

◆ Turn the gas control knob from PILOT to ON.

◆ Set the thermostat between 120°F and 130°F (or, on some models, just above WARM). This is a moderate setting that lowers heating costs, prolongs tank life, and reduces the risk of being scalded.

◆ Replace the access panels.

SMALL JOBS THAT MAKE A DIFFERENCE

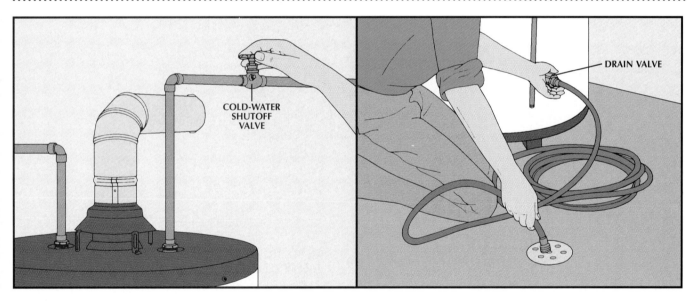

Draining the water heater.

Every few months, empty about 3 gallons of water from the tank to prevent sediment from accumulating. For some repairs, you may also need to drain the tank completely—a process that can take as much as an hour.

◆ Turn off the heat for either type of water heater: In the case of a gas-fired heater, turn the gas control knob off and close the gas shutoff; for an electric model, shut off the power and verify that it is turned off (page 100).

◆ Close the cold-water shutoff (above, left) and open a hot-water faucet in the house to speed draining.

◆ Attach a hose to the drain valve and run it to a nearby floor drain (above, right), an outdoor drain, or a bucket. If you use a bucket, watch carefully to make sure it does not overflow.

◆ As the tank empties, the valve may clog; in that case, open the cold-water shutoff for a few minutes to build up pressure. If the problem persists, stick a screwdriver or a coat hanger into the open valve.

◆ Refill the tank by closing the drain valve and opening the water shutoff. Open the hot-water faucet farthest from the tank; when water flows from that faucet, the tank is full.

◆ Restore power to an electric water heater; turn on the gas and relight the pilot of a gas-fired heater as described at left and above.

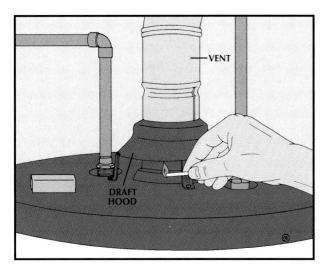

Testing the vent.

Check the venting of a gas-fired water heater once a year to ensure proper ventilation and prevent the backup of dangerous carbon monoxide fumes. Wait until the burner has been on for 5 to 10 minutes, then hold a lighted match under the draft hood *(left)*. If the vent is working properly, the match flame will be drawn in under the edge of the hood. If the flame is blown away from the draft hood or snuffed out, the vent may be blocked or corroded.

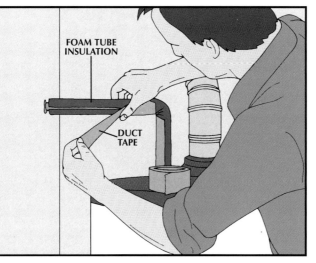

Insulating pipes.

To conserve the energy needed to heat more water, insulate runs of hot-water pipe that pass through unheated areas with adhesive-backed foam, fiberglass tape, or preslit foam tubes secured with duct tape *(left)*.

REPLACING THE THERMOCOUPLE

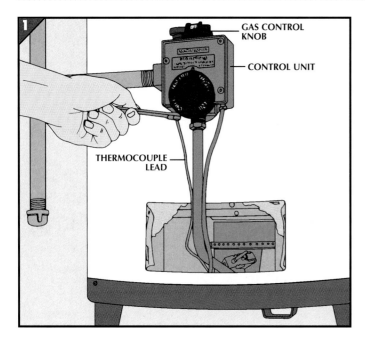

1. Disconnecting the thermocouple.

◆ Turn the gas control knob to OFF and close the gas shutoff valve. With an open-ended wrench, loosen the nut that secures the thermocouple to the control unit *(left)*, then unscrew it by hand.

◆ Pull down on the copper lead to detach the end of the thermocouple from the control unit.

◆ If a second nut secures the thermocouple tip to the bracket, unscrew the nut and slide it back along the copper lead.

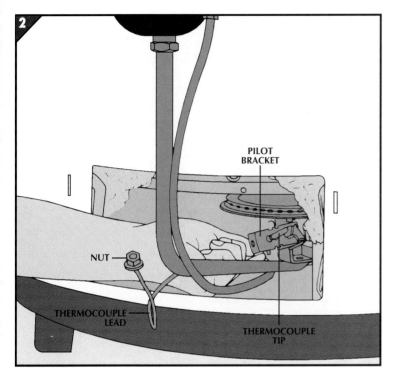

2. Removing the thermocouple.

◆ Grip the base of the thermocouple *(left)* and pull firmly to slide it out of the pilot bracket.
◆ Purchase an exact replacement at a plumbing or heating-supply store.
◆ Push the tip of the new thermocouple into the pilot bracket as far as it will go. If there is a hexagonal nut at the tip, tighten the nut to secure the thermocouple to the bracket.
◆ Run the lead out and then bend it upward in a gentle curve.
◆ Screw the nut on the end of the thermocouple to the control unit by hand, then give it a quarter-turn with an open-ended wrench.
◆ Turn on the gas shutoff valve, wait a few minutes, and relight the pilot *(pages 94-95)*. If it does not stay lit, close the gas shutoff valve and call for service.

DEALING WITH A FAULTY RELIEF VALVE

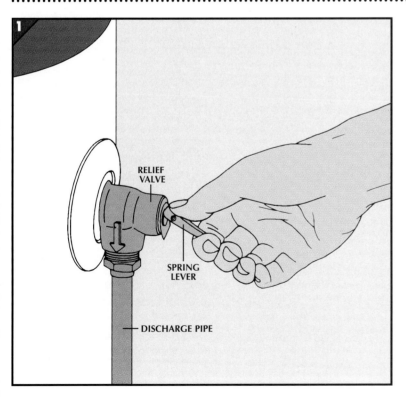

1. Testing the valve.

In the unlikely event temperature or water pressure rises too high inside the water heater, the relief valve opens to prevent the tank from exploding. Check the condition of the valve as follows:
◆ Place a bucket under the discharge pipe and lift the spring lever *(left)*, keeping away from the outlet of the discharge pipe as hot water escapes. About 1 cup of water should spurt out. If no water comes out, proceed to Step 2 to replace the valve.
◆ Otherwise, lift the lever several times to clear the valve of sediment.
◆ If water continues to drip, replace the valve *(Steps 2 and 3)*.

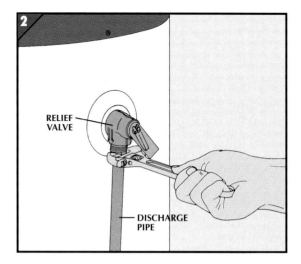

2. Removing the relief assembly.

◆ Turn off the heat: For a gas-fired water heater, turn the gas control knob off and close the gas shutoff; for an electric model, shut off the power and verify it is off *(page 100)*.

◆ Close the cold-water shutoff.

◆ If the relief valve is on the side of the tank, drain about 5 gallons of water from the tank *(page 95)*. For a valve that is on the top of the tank, drain 1 gallon of water.

◆ Loosen and remove the discharge pipe, which may be soldered to a threaded adapter or screwed into the valve *(left)*.

◆ Unscrew the old valve from the tank with a pipe wrench. In an old tank, the valve may be difficult to remove. Have a helper brace the tank if necessary and use firm, steady pressure on the wrench. Do not jerk the valve; you could damage the tank.

3. Installing a new relief valve.

◆ Take the old valve with you to buy an exact replacement. The model type and size often appear on a metal tag hanging from the valve.

◆ Wrap the threads of the new relief valve with plumbing-sealant tape. Screw it into the tank by hand, then tighten it with a pipe wrench. Screw the discharge pipe—or the pipe's threaded adapter—into the valve outlet.

◆ Refill the water heater. Open the gas shutoff and relight the pilot *(pages 94-95)* of a gas-fired water heater; turn the power to an electric one back on.

◆ If the valve continues to leak, check the house water pressure *(page 44)*.

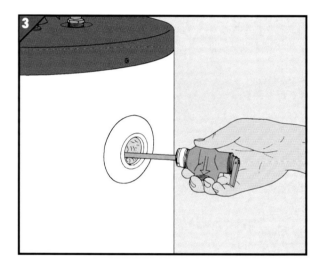

SERVICING THE DRAIN VALVE

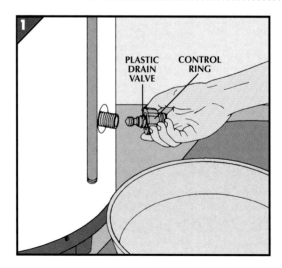

1. Repairing a leaking drain valve.

◆ Turn off the heat: For a gas-fired water heater, turn the gas control knob off and close the gas shut-off; with an electric water heater, turn off power and confirm it is off *(page 100)*.

◆ Close the cold-water shutoff valve and drain the heater completely *(page 95)*.

◆ If the drain valve has a removable handle, unscrew it and replace the washer behind it. Reassemble the handle and fill the tank. If the drain valve still leaks, empty the tank and remove the valve by turning it at the base with a pipe wrench.

◆ Also remove a leaking metal valve with a handle that does not come off, or a leaking plastic valve *(left)*. To take off a plastic valve, turn the control ring counterclockwise by hand, four complete revolutions. Then, while pulling firmly on the handle, turn the ring clockwise six complete turns.

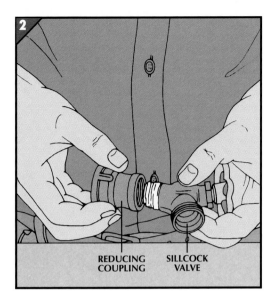

REDUCING COUPLING | SILLCOCK VALVE

2. Assembling the new drain valve.

Replace the valve you removed—whether it was plastic or metal—with a durable metal sillcock. The new valve is likely to require a $\frac{3}{4}$- to $\frac{1}{2}$-inch reducing coupling in order to fit it to the tank. Wrap plumbing-sealant tape around the threaded end of the sillcock and screw it into the $\frac{1}{2}$-inch end of the reducing coupling (left).

3. Installing the new valve.

◆ Apply plumbing-sealant tape to the threaded nipple at the tank's drain outlet.
◆ Screw the coupling and sillcock onto the nipple and tighten by hand. Tighten the coupling to the nipple with a pipe wrench (right), turning another quarter-turn or so.
◆ To align the valve, fit an adjustable wrench over the body of the sillcock valve (not over its outlet). Turn clockwise until the valve faces the floor.
◆ Refill the tank. Then either relight the pilot (pages 94-95) or turn the electricity back on, depending on the type of water heater.

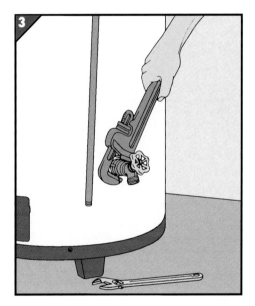

TRICKS OF THE TRADE

Swapping Valves without Draining the Tank

When replacing a drain valve, you can avoid the trouble of emptying the tank. Before leaving the house for work, turn off the water heater. On your return, check to be sure the water has cooled. Prepare the new valve and place a pan and towel under the old one. Close the cold-water shutoff and all household faucets. Begin draining water from the valve with a hose; after about 10 minutes, the flow will stop as a vacuum forms in the tank, although small amounts of water may spurt out. Remove the old valve and quickly insert and tighten the new one. Refill the tank and turn the gas or electricity back on.

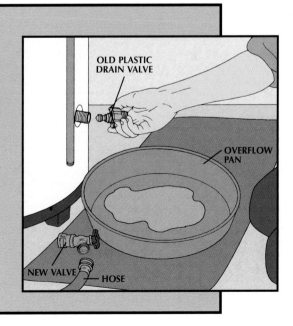

OLD PLASTIC DRAIN VALVE

OVERFLOW PAN

NEW VALVE | HOSE

TESTING ELECTRIC HEATING ELEMENTS

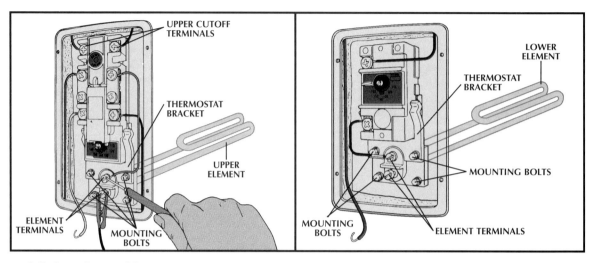

Finding and fixing the problem.

◆ Turn off power at the service panel.
◆ To double-check at the heater that the power is off, remove the cover of the upper control panel. Push aside the insulation and touch the probes of a voltage tester *(below)* to the two upper terminals of the high-temperature cutoff—the top ones on the panel. If the tester does not light, the power is off.
◆ Detach a wire from one of the element terminals. Clip the lead of a continuity tester to a terminal and touch the probe to the other one *(above, left)*. If the tester lights up, the element

is good. Reattach the wire.
◆ Remove the lower control panel's cover and test the lower heating element the same way *(above, right)*.

If both pass the continuity test, call a technician to repair the heater. Replace an element that fails the test as follows:
◆ Drain the heater *(page 95)*.
◆ Remove the element's mounting bolts, set aside the thermostat bracket, and pull out the element. An element that has no bolts can be unscrewed with a socket wrench. Scrape scale and rust from around the tank opening.

◆ Purchase a matching element, or a replacement that is the same voltage and no higher in wattage.
◆ Slide a new gasket onto the element and insert it in the heater. Position the lower mounting bolts, then replace the thermostat bracket and put in the two upper bolts; tighten each of the bolts a little at a time to ensure even pressure on the gasket. Install a screw-in element with a socket wrench.
◆ Reconnect the terminal wires, repack the insulation, and replace the access panels. Fill the heater with water and turn the power back on.

INSTALLING AN ELECTRIC WATER HEATER

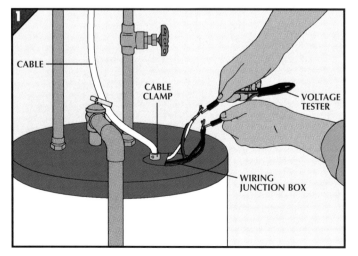

1. Disconnecting the wiring.

◆ Turn off power at the service panel.
◆ To confirm power is off, remove the cover of the heater's wiring junction box. Touching only the wire insulation, pull the wires out. Unscrew the wire caps without touching the bare wires. Touch a voltage tester's probes *(left)* to the twisted wire pairs. Repeat the test between each pair and the ground wire, which is attached to a screw inside the box. The tester will not light if the power is off.
◆ Draw a diagram of the wiring arrangement for reference, then untwist the wires. Disconnect the ground, loosen the clamp, and pull the cable free.

If the cable wires are aluminum, plan to have an electrician perform the final hookup *(Step 7)*.

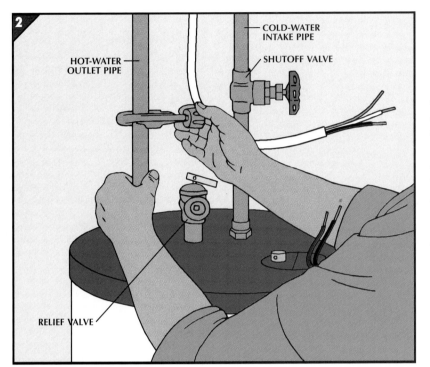

2. Removing the pipes.

◆ Turn off the cold-water shutoff valve and drain the heater *(page 95)*. Turn on a hot-water faucet to make sure the hot-water supply lines are empty.

◆ Disconnect the discharge pipe from the temperature-pressure relief valve; depending on how it is attached, this may require desoldering the pipe from a threaded adapter. If the discharge pipe is in good condition and will fit the relief valve on the new water heater, you can reuse it if it will extend to within 6 inches of the floor when installed.

◆ Cut the hot-water outlet pipe about 8 inches above the tank *(left)*. Cut the cold-water intake pipe at least 3 inches below the shutoff valve and, if possible, about 8 inches above the tank.

3. Removing the old heater.

◆ Grasp the hot- and cold-water pipe stubs at the top of the tank and gently shift the heater back and forth until it is clear of the cut pipe ends.

◆ Tilt the heater toward you and walk it out of the work area, using the pipe stubs as handles.

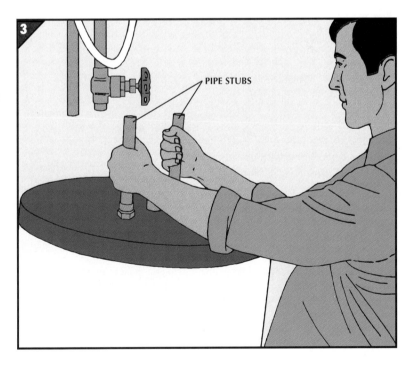

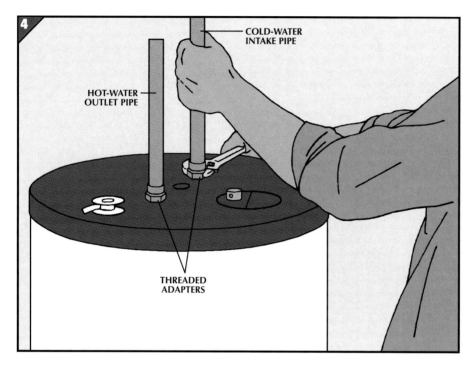

4. Attaching pipes to the new tank.

◆ Cut two lengths of $\frac{3}{4}$-inch copper pipe, each about 1 foot long.

◆ Sweat a $\frac{3}{4}$-inch threaded adapter onto one end of each *(pages 17-18)*.

◆ When the work has cooled, wrap plumbing-sealant tape around the threads and screw one of the pipes by hand into the cold-water intake, the other into the hot-water outlet.

◆ Tighten the connections with a wrench *(left)*.

5. Installing the relief valve.

◆ Walk the new water heater into position as you removed the old one *(Step 3)*, making sure to align the outlet pipe with the hot-water line and the intake pipe with the cold-water line and shutoff.

◆ Wrap the threads of a new relief valve with plumbing-sealant tape and screw the valve into place *(right)*. Tighten the valve with a pipe wrench so that the discharge pipe will not run into the hot- or cold-water pipes or block access to control-panel covers. For a valve on the side of a tank, point the pipe opening straight down.

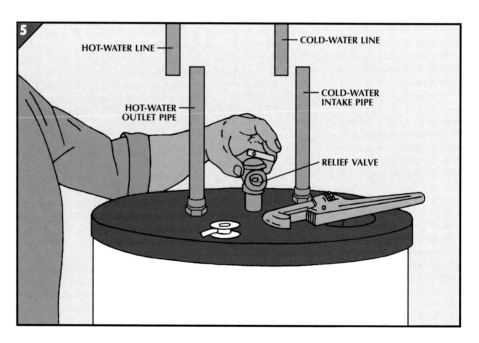

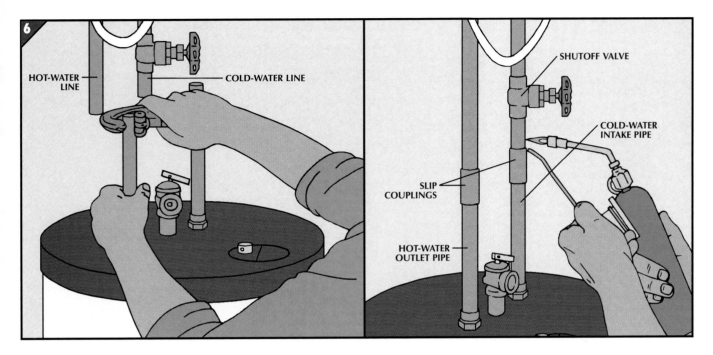

HOT-WATER LINE

COLD-WATER LINE

SHUTOFF VALVE

COLD-WATER INTAKE PIPE

SLIP COUPLINGS

HOT-WATER OUTLET PIPE

6. Connecting the plumbing.

◆ Trim both pipes rising from the top of the water heater about $\frac{1}{4}$ inch below the ends of the corresponding house pipes *(above, left)*.
◆ Prepare two slip couplings for soldering and slide one onto each pipe. Shift the tank to align the pipes.
◆ Position each coupling and solder in place *(above, right)*,

using the methods described on pages 17 and 18.
◆ Attach a discharge pipe to the relief valve.
◆ Make sure all household faucets are closed, then open the cold-water shutoff. If a joint leaks, you will hear air hissing from it; close the valve and resweat the joint.
◆ Reopen the valve and fill the tank. Open an upper floor's hot-water faucet. When water comes out, the tank is full.

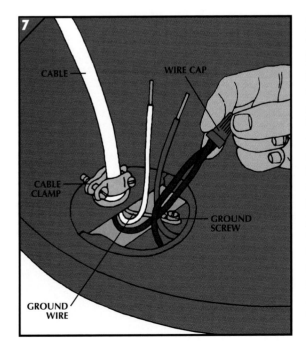

CABLE

WIRE CAP

CABLE CLAMP

GROUND SCREW

GROUND WIRE

7. Connecting the wires.

When wires inside the house cable are aluminum, have an electrician connect them. Otherwise, proceed as follows:
◆ Remove the cover of the wiring junction box and pull the two heater wires out. Pass the cable through the clamp and tighten it with a screwdriver.
◆ Join the heater and cable wires, consulting the diagram prepared in Step 1; commonly, black connects to black and red to white. Twist each wire pair and secure the connection with a wire cap *(left)*. Attach the cable ground wire to the ground screw in the junction box.
◆ Open the control panels *(page 100)* and check that the thermostats are at the correct temperature. Thermostats come preset at 120°F, which should be adequate for most homes. Do not set the thermostat above 140°F.
◆ Close the panels; turn on power at the circuit panel.

Bringing Water Outdoors

4

Adding a backyard hydrant, a sillcock to feed a garden hose, a shower, or other outdoor amenities involves pipe arrangements that are not very different from those for indoor plumbing. But as this chapter explains, there are a few necessary adjustments; most notably, external supply lines must be planned to avoid their freezing in winter.

A pop-up sprinkler head →

Extending water lines outside your house can pay rich dividends in convenience: A new line may serve a sillcock on an exterior wall *(opposite)*, a yard hydrant near a flower garden *(page 112)*, or an outdoor shower *(pages 113-115)*. Most such additions call for no more than basic pipe fitting skills, but a shower may require professional excavation, because the shower water must drain into the sewer system. Have the contractor shore up the sides of the trench as shown on page 115.

Pipe Material: Polyethylene (PE) pipe and polyvinyl chloride (PVC) pipe are both good choices for outdoor extensions of a plumbing system: They are inexpensive, durable, and easy to work with. PE pipe, which is flexible and sold in long rolls, may require fewer fittings than rigid PVC pipe; however, its fittings are inserted into the pipe and will reduce water flow somewhat.

Because both types of plastic pipe deteriorate in direct sunlight, any portion of the run that is aboveground should be covered by two coats of latex paint or enclosed.

Tapping the Household Supply: As the water source for an outdoor line, select a supply pipe close to the extension's exit point from the house—preferably a pipe in the basement, minimizing problems of access. If the extension will have a run of more than 300 feet or will serve numerous outlets, use 1-inch pipe. Otherwise, match the extension to the size of the indoor pipe— most likely $\frac{1}{2}$ or $\frac{3}{4}$ inch.

Cold-Weather Safeguards: Some outdoor fittings are specifically designed to cope with freezing temperatures. In the absence of such protection, however, outdoor plumbing may burst in cold weather unless you include a valve that allows water to be drained out before winter arrives. In many cases, this valve can be located in the basement *(pages 108 and 110)*. If such a handy arrangement is not possible, you will need an exterior valve that permits the extension pipe to be flushed by air pressure *(page 111)*.

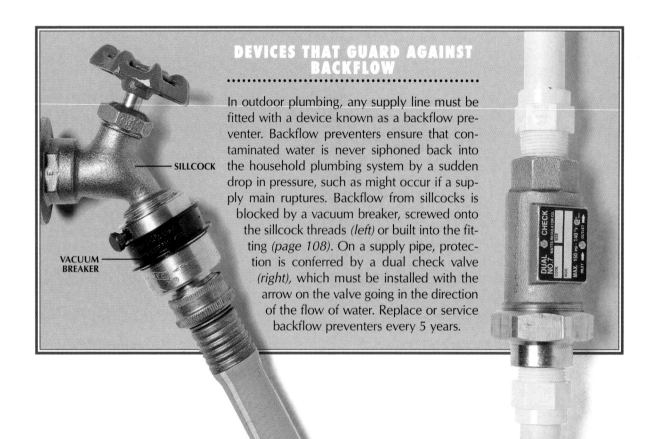

DEVICES THAT GUARD AGAINST BACKFLOW

In outdoor plumbing, any supply line must be fitted with a device known as a backflow preventer. Backflow preventers ensure that contaminated water is never siphoned back into the household plumbing system by a sudden drop in pressure, such as might occur if a supply main ruptures. Backflow from sillcocks is blocked by a vacuum breaker, screwed onto the sillcock threads *(left)* or built into the fitting *(page 108)*. On a supply pipe, protection is conferred by a dual check valve *(right)*, which must be installed with the arrow on the valve going in the direction of the flow of water. Replace or service backflow preventers every 5 years.

SILLCOCK

VACUUM BREAKER

TOOLS

Drill
Spade drill bit
Star drill
Ball-peen hammer
Screwdriver
Hammer
Shovel
Plastic tarpaulins
Garden hose

Carpenter's level
Pipe wrench
Pipe cutter
Sledgehammer
Adjustable wrench
Measuring stick

MATERIALS

Silicone caulk
Plumbing-sealant
   tape
Plastic pipe primer
   and cement
Wooden stakes
String
Swimming pool
   patching material

Tar
Gravel
Plywood
2 x 4 lumber
Newspaper

SAFETY TIPS

Wear gloves and safety goggles to solder copper supply pipe. When digging, gloves and a back brace help prevent injury. Gloves, goggles, and a hard hat are necessary if you must cut a sewer line in a deep ditch.

INSTALLING A SILLCOCK

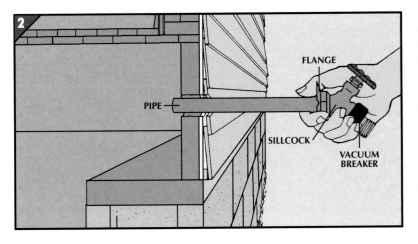

1. Boring a hole for the pipe.

For an ordinary sillcock, the supply pipe should pass through the wall horizontally. For a freezeproof sillcock *(page 108)*, the hole should slope slightly downward toward the outside.

◆ On a wood wall, bore the hole just above the masonry foundation with a spade bit and, if necessary, an extension. Make the hole just large enough for the pipe.

◆ For masonry, use a star drill. Set the point on a horizontal strip of mortar or at the center of a cinder block, and drive the drill with a ball-peen hammer, turning the point a few degrees between blows. Make the hole diameter 1 inch larger than the pipe to accommodate a sleeve *(pages 110-111)*.

FLANGE

PIPE

SILLCOCK

VACUUM
BREAKER

2. Assembling the sillcock.

◆ Screw a vacuum breaker onto the sillcock. Attach the sillcock to a pipe 3 inches longer than the thickness of the wall, using an adapter *(page 13)*, if necessary. A freezeproof sillcock does not require a separate pipe, but must have a stem at least 2 inches longer than the wall thickness.

◆ For masonry walls, run the pipe through a sleeve *(pages 110-111)* to prevent corrosion and damage to the pipe.

◆ Insert the assembly into the hole and screw the flange to the wall, using anchors, if necessary. Fill any gaps with silicone caulk.

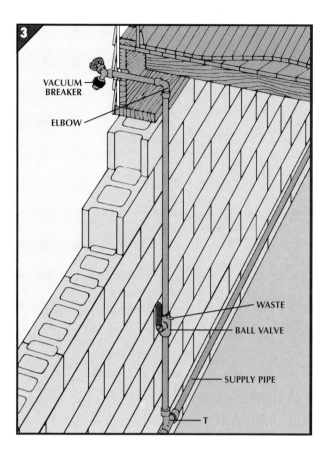

VACUUM BREAKER

ELBOW

WASTE

BALL VALVE

SUPPLY PIPE

T

3. Tapping a basement supply line.

◆ Insert a T fitting in a convenient supply pipe, add a short section of pipe, and install a ball valve *(page 45)* with a waste, orienting the valve so that the waste plug is on the side nearest the sillcock.

◆ Attach an elbow on the end of the sillcock pipe, and run a section of pipe from the valve to the elbow.

◆ Finish the job by caulking any gaps around the hole on the basement side of the wall.

◆ To drain the line, close the valve, open the outdoor faucet, and then open the valve's waste plug.

A FREEZEPROOF SILLCOCK

A freezeproof sillcock has all the parts of a standard stem faucet, but its elongated body allows the stem to stop the flow of water inside the house, where the temperature is above freezing. Because the sillcock is installed sloping slightly downward toward the outside, water drains out when the valve is closed. Choose a model with a built-in vacuum breaker: adding a screw-on breaker could prevent proper draining.

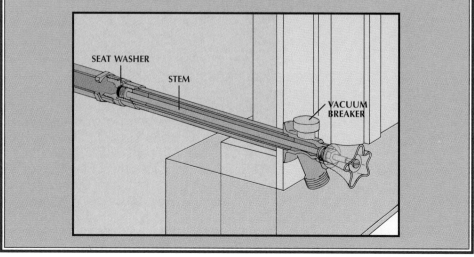

SEAT WASHER

STEM

VACUUM BREAKER

LOCATION OF NEW OUTLET

1. Staking out a trench.

◆ Drive a stake into the ground at the location for the yard outlet and another stake at the wall where the supply pipe will exit the house. Tie a string between the two stakes.

◆ With a flat shovel, make two parallel grooves about 3 inches to the left and the right of the string, pushing the blade down about 2 inches into the ground.

◆ Remove the string and stakes. Use the shovel to divide the sod between the grooves into segments 5 or 6 feet long. Spread a plastic sheet next to one of the grooves.

◆ Push the shovel under the sod at the corner of a segment and work it up and down to free the roots. Repeat along the rest of the segment, then lift the sod carpet and place it on the plastic sheet.

2. Digging the trench.

◆ Spread a large plastic sheet or a tarpaulin on the ground.

◆ Working in from the sides of the cleared area, dig a V-shaped trench about a foot deep, piling the loose soil on the plastic sheet.

◆ For a freezeproof hydrant *(page 112),* dig the trench to a depth below the frostline in your area.

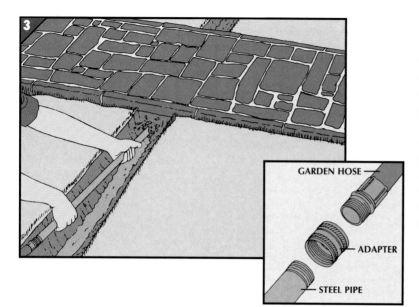

3. Tunneling under a paved walk.

◆ Have a 3-foot length of $\frac{3}{4}$-inch steel pipe threaded at one end, and join it to a garden hose with an adapter *(inset)*.

◆ Turn the water on full force and stick the end of the steel pipe into the ground under the walk, using the pipe as a water-pressure pick to loosen the earth. The stream of water should be strong enough to keep the open end of the pipe clear and to sweep away loose earth.

◆ When you reach the middle of the walk, cross to the other side and work from there to complete the tunnel.

GARDEN HOSE

ADAPTER

STEEL PIPE

TWO TACTICS FOR WINTERTIME DRAINING

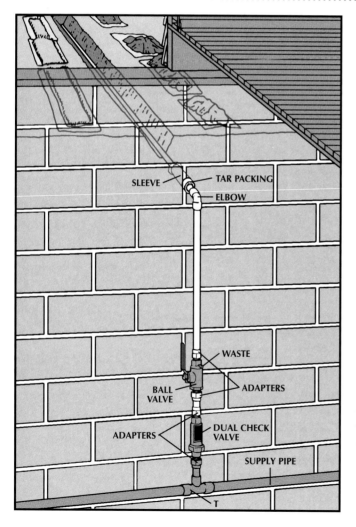

SLEEVE — TAR PACKING
— ELBOW

WASTE

BALL VALVE — ADAPTERS

ADAPTERS — DUAL CHECK VALVE

SUPPLY PIPE

T

A drain in the basement.

If the ground outside your house is level or slopes slightly downward toward the house, you can tap water for a yard hydrant from a basement supply line and also drain the system from there.

◆ Bore a hole through the wall *(page 107)* at the bottom of the trench, sloping it slightly downward into the basement.

◆ Insert a T fitting in the supply pipe, followed by a short length of pipe.

◆ Install a dual check backflow preventer *(page 106),* add another short length of pipe, and install a ball valve *(page 45)* with a waste, orienting the valve with the waste plug toward the outside.

◆ Run a pipe from the valve to an elbow.

◆ Make a sleeve from a PVC pipe 1 inch larger in diameter than the supply pipe and slightly longer than the wall thickness. Insert the sleeve in the exit hole.

◆ Run pipe from the end of the trench through the sleeve. With a carpenter's level, make sure that the pipe is pitched slightly downward toward the house along its entire length. If necessary, give it the proper pitch by propping it up with fragments of brick or stone at 6-foot intervals, then shoring up the pipe between the fragments with loose earth.

◆ Attach the pipe to the elbow.

◆ Seal the sleeve by packing the space around the pipe with tar. Alternatively, slip rubber bushings over both ends of the sleeve before inserting the pipe, then clamp the bushings to both sleeve and pipe.

A Watertight Connection through a Wall

To seal an underground hole through masonry, fill gaps around the sleeve with the material used to patch concrete swimming pools. Made predominantly of Portland cement, this material doesn't shrink when it dries and can solidify even under water. It is generally available at swimming pool supply stores.

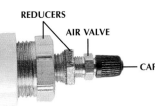

A valve to drain a line with air.

If there is no basement or the lawn slopes away from the house, a different supply and drainage arrangement is required for a yard hydrant.

◆ As the water source, choose a supply pipe near an outside wall and install a T, a dual check valve, and a ball valve as on page 110.

◆ Bore a hole in the wall *(page 107),* run pipe from the valve through the wall, and attach an elbow followed by a short length of pipe.

◆ Add a T with a threaded outlet, then extend another section of pipe to the bottom of the trench.

◆ Attach an elbow and run pipe from the elbow to the yard hydrant.

◆ Screw reducers onto the threaded T until the opening fits a $\frac{1}{8}$-inch air stem valve *(inset).*

◆ To drain the pipe for winter, hook a bicycle pump or an air compressor to the air valve and flush the water out through the open yard outlet, keeping the pressure below 60 pounds.

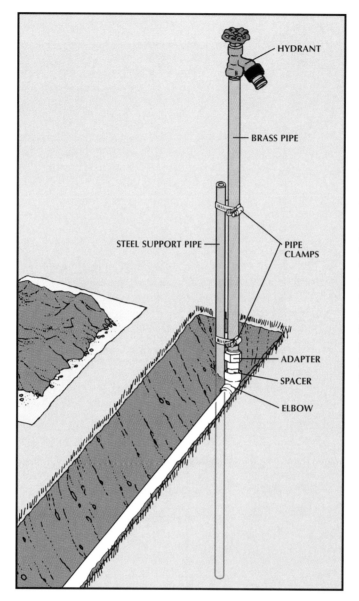

Bracing the hydrant.

◆ Cut a 40-inch-long support from a $\frac{3}{4}$-inch steel pipe. At the far end of the pipe in the trench, drive this steel pipe about 2 feet into the earth with a sledgehammer.

◆ For the vertical riser, use a brass pipe threaded at both ends. Attach an elbow to the end of the pipe in the trench, and add a short spacer. Using plumbing-sealant tape, attach a threaded adapter to one end of the brass pipe, then glue the adapter to the spacer on the elbow.

◆ Strap the two vertical pipes together with stainless-steel pipe clamps and turn on the water briefly to flush dirt from the supply pipe.

◆ Wrap plumbing-sealant tape around the threads of the brass pipe and screw the hydrant to the pipe. Turn on the water to test for leaks before filling the trench.

FREEZEPROOF YARD HYDRANTS

Freezeproof yard hydrants—functional year-round—come in two basic types: ground-level *(left)* and freestanding *(right)*. Both have a valve that lets water drain into the ground from the stem when the hydrant closes. To install a freezeproof hydrant, dig the trench for the supply pipe deeper than the frostline in your area. At the end of the trench, dig a drainage pit at least 8 inches deeper than the trench and add 8 inches of gravel at the bottom. Connect the supply pipe to the freezeproof hydrant in the same manner for an ordinary yard hydrant *(above)*.

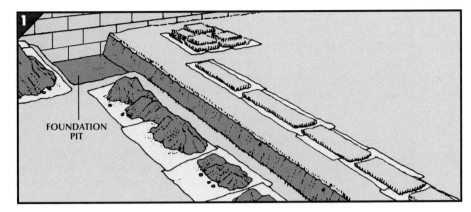

1. Digging a foundation pit.
◆ Choose a site at least 7 feet away from doors or windows and dig a pit 1 foot deep and about 6 inches larger all around than the shower base.
◆ Mark out and dig a trench *(page 109)* from one side of the foundation pit to the area of the main sewer line *(page 115)*.

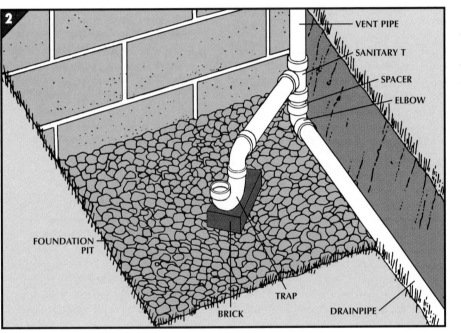

2. Installing the drain and vent.
◆ Fill the pit with coarse gravel a few inches deep and mark the center of the shower base with a brick.
◆ Set a trap with the same diameter as the shower-base drainpipe on the brick. Run pipe from the trap to the pit's back corner on the trench side.
◆ Cement a sanitary T *(page 14)* on the end of the drainpipe, attach a short spacer pipe to the bottom of the T, and add an elbow to the spacer, positioned so the outlet of the elbow faces the trench.
◆ Run pipe from the elbow out into the trench and on to the sewer line.
◆ Attach a vent pipe to the top of the sanitary T extending at least 7 feet high and above the highest window of the house.

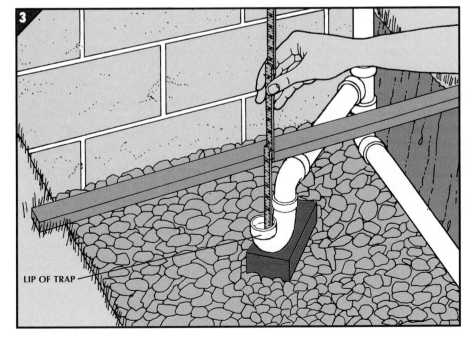

3. Installing the vertical drainpipe.
◆ Place a straight piece of wood across the top of the gravel pit, and measure the distance from the bottom of the lip of the trap to the bottom of the wood.
◆ Remove the drain strainer from the shower base and measure the depth of the drain hole.
◆ Add the two measurements, subtract $\frac{1}{2}$ inch, and cut a length of drainpipe equal to the total.
◆ Install the drainpipe in the trap *(page 42)* and cover it temporarily to prevent objects from falling in.

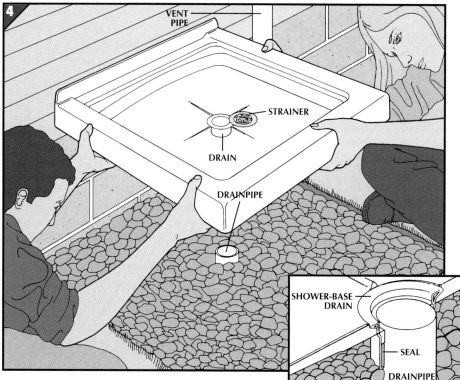

4. Installing the shower base.

◆ Fill the trench next to the pit with loose earth and replace the sod cover. Fill the foundation pit with gravel to ground level and tamp the gravel down with a 2-by-4.

◆ Uncover the drain and, with a helper, lower the shower base carefully onto the exposed end of the drainpipe.

◆ When the base is firmly in place, place the seal that comes with the shower kit around the drainpipe *(inset),* or fill the space with silicone caulk, and replace the strainer.

◆ Replace dirt and sod around the edge of the shower base and vent pipe.

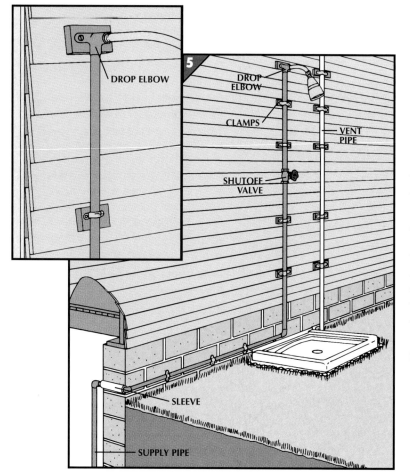

5. Bringing water to the shower.

◆ Tap a supply pipe and add a valve with a waste *(page 108).* Run a pipe along the house wall to a point directly above the center of the shower base, then straight up about 4 feet.

◆ Add a shutoff valve and about 2 more feet of vertical pipe topped by a drop elbow with a threaded inlet *(inset).*

◆ Fasten the drop elbow to the wall with screws, and clamp the supply and vent pipes to the wall at 2-foot intervals. If the house has wood siding *(inset),* nail wood blocks between the siding and the drop elbow and pipes, and fasten the elbow and clamps to the blocks with screws.

◆ Screw a shower arm and a shower head into the drop elbow.

◆ Cover the shower when not in use to keep rainwater out of the septic tank or sewer. Freeze-proof it for winter by pouring nontoxic antifreeze *(page 38)* in the trap and covering it securely.

PLYWOOD

WOOD BRACES

SEWER PIPE

6. Exposing the sewer line.

Almost all building codes require that the gray-water waste from a shower be drained into the household sewer system.

◆ Stake out the location of the main sewer line in your yard. Hire a contractor with a backhoe to dig a trench to within a foot of the pipe, reinforcing the walls of the trench with plywood braced with 2-by-4s. For clarity, some braces have been omitted from the foreground of this drawing.

◆ Dig around the section of pipe you will be tapping so that it is completely exposed.

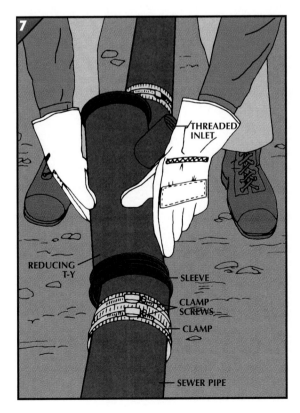

THREADED INLET

REDUCING T-Y

SLEEVE

CLAMP SCREWS

CLAMP

SEWER PIPE

7. Tapping into the sewer line.

◆ To tap into cast-iron pipe, use the method described on page 25, but install a reducing T-Y *(page 14)* with a threaded inlet the same size as the shower drainpipe.

◆ After clamping the reducing T-Y in place, wrap plumbing-sealant tape around the threads of a plastic adapter and screw it into the threaded inlet of the reducing T-Y.

◆ Connect the shower drainpipe to the adapter.

For plastic sewer pipe, install a reducing T-Y with an inlet the same size as the shower drainpipe. Connect the shower drainpipe to the inlet of the reducing T-Y.

⚠ *Sewer gas is toxic. When tapping a sewer line, temporarily plug the open pipes with*
CAUTION *newspaper.*

An Underground Sprinkler System

Few homeowners enjoy dragging a sprinkler around the yard to keep the lawn green during summer dry spells. Fortunately, a fully automated alternative is available—a sprinkler system buried in the ground and controlled from a timing box in the basement. In response to the timer's commands, electrically operated valves outside the house open, and water flowing through the system causes sprinkler heads to push up through the grass and create overlapping umbrellas of spray. At the end of the watering period, the heads sink back into the ground. If you want to deviate from your preset watering program, you can manually override the system's automatic controls.

A Tailor-Made Plan: Most distributors and manufacturers of sprinkler parts will supply you with a blueprint of a system tailored to your home in exchange for your promise to buy their components. The designer will need a rough scale map of your house and grounds, indicating where you have grass and where you have flowers or bushes; which areas are sunlit and which shady; the location of large trees or other obstructions; and the nature of your soil— whether sandy, rocky, or compact. In order to recommend the appropriate type and size of pipe, the designer must also know the water pressure in your supply lines; measure the pressure as explained on page 44. Rigid PVC pipe is the usual choice for a sprinkler system, although polyethylene (PE) pipe may be preferred for some situations *(page 121)*.

Winterizing: In cold climates, all exterior pipes must be drained to prevent their bursting. Typically, two special valves are required— one outdoors for flushing the sprinkler network by air pressure, and one inside the house to drain water from the supply line that leads to the control valves.

 **TOOLS**

Straight-edged shovel
Hacksaw or PVC cutting tool

 **MATERIALS**

Graph paper for scale map
Stakes and string
Sprinkler heads
Control valves with anti-backflow protection
Sprinkler timer
PVC pipes and fittings

A three-zone sprinkler system.
No two sprinkler systems are identical, but the one shown at right illustrates common situations: a sunny front lawn, a tree-shaded side lawn, and a flower garden. Separate supply lines feed the three networks of pipe, called circuits, that serve the watering zones; electrical control valves (indicated by numbered squares) are clustered near where the main supply pipe exits from the basement. The pipe layout between the control valves and sprinkler heads is shown in green, yellow, and blue, with the pipe diameters indicated for each run. The flower garden at the top is watered by simple pop-up spray heads (triangles), the two lawns by impulse pop-ups (circles). When in operation, these heads spray water in the patterns indicated by large half and quarter circles. Each pattern overlaps other patterns by about 60 percent to provide a more uniform depth of coverage. (The outer rim of a head's area of coverage gets less water than the ground closest to the head.)

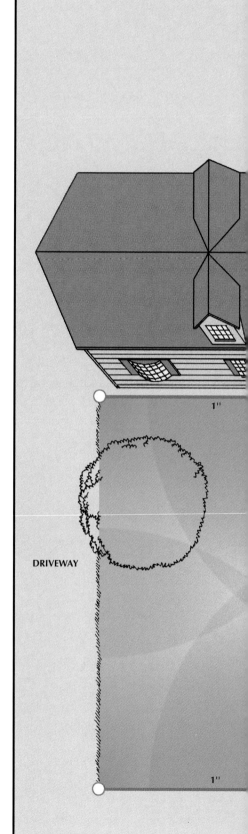

DRIVEWAY

116

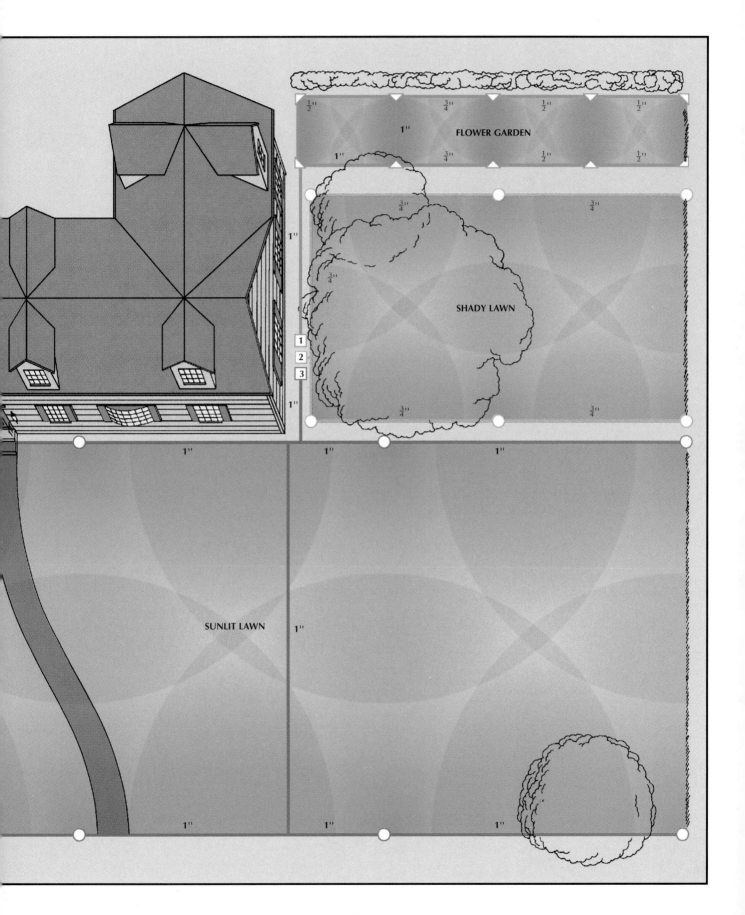

FLOWER GARDEN

SHADY LAWN

SUNLIT LAWN

POP-UP SPRAY

PLASTIC POP-UP IMPULSE

BRASS POP-UP IMPULSE

BRASS STANDING IMPULSE **PLASTIC STANDING IMPULSE**

A gallery of sprinkler heads.
Spray sprinkler heads are used to send a steady shower of water over a small-to-medium area. Impulse heads shoot out long jets of water from a rotating nozzle on medium-to-large areas. Both are available with working parts of brass or plastic and come in "pop-up" or "standing" versions. A pop-up head serves for a traveled area. When the control valve is opened, water pressure forces a piston in the sprinkler head upward; gravity or a spring within the head brings it down again when the water is shut off. A standing version is permanently raised aboveground in an out-of-the-way location—along the perimeter of the yard, for example.

INSTALLING THE CONTROLS

1. Connecting the control valves.
◆ Tap a basement supply line and fit the extension pipe with a ball valve with a waste *(page 110)*. Drill the extension pipe's exit hole so it has a slight inward slope. Run the control-valve wires and pipe through the hole.
◆ Install an elbow on the extension pipe *(right)*, then attach a vertical pipe and another elbow. Add a pipe with a T attached. Add a similar length and T for a two-circuit system, and so on.
◆ Clamp the assembly to the house, sloping it slightly toward the elbow.
◆ Into each T, insert a riser long enough to raise the control valve at least 6 inches above the ground. Install the valve on the riser.
◆ In cold climates, attach a $\frac{1}{8}$-inch air stem valve on the vertical pipe *(page 111)*. Before winter, close the ball valve in the basement and use an air compressor at the air stem valve to flush water out of the sprinkler networks. Open the waste on the ball valve to drain remaining water.

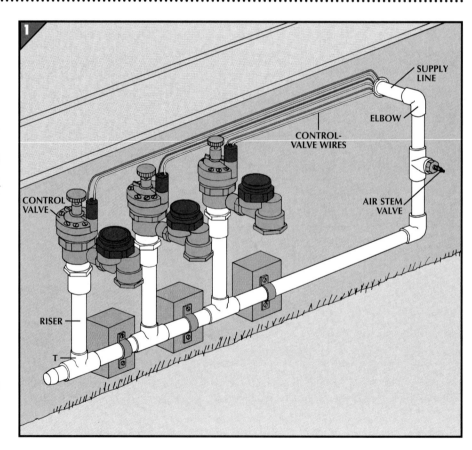

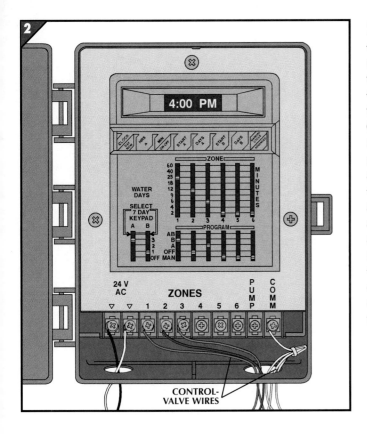

2. Wiring the timer.

◆ Affix the control box to the basement wall near the point where the control-valve wires enter from the outside.

◆ Thread either of the two wires from each valve through the right-hand hole below the terminals. Connect the valve wires to the terminals by matching the terminal and valve numbers. (The power line for the control box should already be attached to the terminals marked 24 V AC.)

◆ Thread the remaining valve wires into the box and, using a wire cap, join them to each other and to a short length of wire of the same kind and size; connect this wire to the terminal marked COMM (common).

◆ Switch the main box control to OFF. Plug the control box cord into an electrical outlet. Set the control box to your desired schedule, according to the manufacturer's specifications.

CONTROL-VALVE WIRES

ADDING PIPES AND SPRINKLER HEADS

1. Running the pipe.

◆ Working with one circuit at a time, drive stakes into the ground where the sprinkler heads will be. Stretch strings in between as guides to dig straight trenches for the pipe *(page 109).*

◆ Assemble straight sections of pipe and install fittings: an elbow for a sprinkler head at the end of a pipe; a T for a head in a straight run of pipe or for the addition of two pipes; and a T-and-elbow combination, linked by a short spacer, for a head at the corner *(inset).*

◆ Set each completed run of pipe into the trench and connect the runs inside the trench. Connect the pipe network to its control valve.

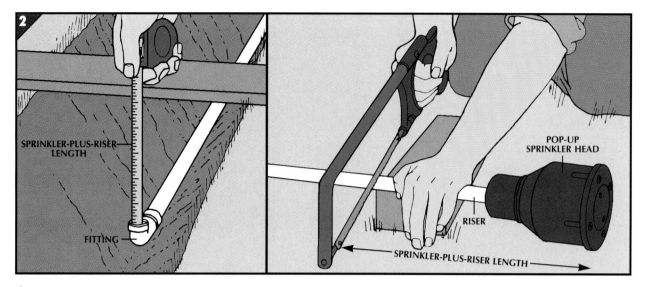

2. Connecting the sprinkler heads.

Pop-up sprinkler heads should be installed so that their tops are flush with the ground.

◆ Set a straight board across the top of the trench over a sprinkler fitting and measure the distance from the board to the inside of the fitting *(above, left)*. This is the length of the sprinkler head with its attached riser.

◆ Attach a sprinkler head to an adapter and a section of pipe. Measuring down from the top of the head, cut the pipe to give the sprinkler-plus-riser length measured *(above, right)*. Repeat this procedure with all other heads.

◆ Finally, flush out the pipes and then install the head assemblies.

3. Testing the system.

◆ With the pipes still exposed in the trenches, open the control valve for a few minutes to check for leaks in the pipes and observe the spray patterns of the heads.

◆ If a head sprays an oval pattern rather than a circle, the riser beneath is not vertical; drive a short vertical stake into the trench next to the riser and, with heavy wire, tie the riser to the stake so that it is tight *(above)*.

◆ If a head sprays in the wrong direction, twist it on the riser to correct the aim.

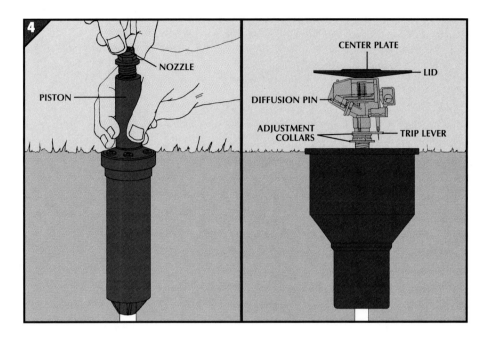

4. Adjusting the spray.

To change the pattern or trajectory of the water from a pop-up spray head *(above, left)*, pull the piston out and hold it with one hand while unscrewing the nozzle with the other; reverse the procedure to install a new nozzle.

With an impulse head *(above, right)*, pry off the center plate, then pull out the cotter pin below to release the entire lid, providing access to the controls below. To water any area less than a full circle, slide the adjustment collars from side to side with the trip lever in the down position. For full-circle coverage, flip the trip lever up. To adjust the distance and dispersion of the spray, screw the diffusion pin in or out of the water stream.

THE POLYETHYLENE OPTION

Flexible polyethylene plastic pipe may be best for a sprinkler system if the yard is irregularly shaped or if the pipe must follow a winding course. Sold in long rolls, it is easy to cut with a sharp knife or a specialized ratchet cutter. (Do not cut with a sawing motion.) Fittings for polyethylene, unlike those for rigid PVC, are inserted inside the pipe and reduce water flow; if you choose PE pipe, buy it one standard pipe size larger than the diameter recommended for PVC. It is better to use PVC pipe to connect the basement supply line to the control valves, since that pipe is under constant pressure.

Household sewage—99 percent of which is water—is treated in the backyard in many suburban and rural homes. Most often, all the wastewater is directed to a septic tank and seepage field *(opposite)*. Where water is scarce, some local codes permit homeowners to divert and reuse so-called gray water— wastewater from clothes washers, tubs, showers, and bathroom sinks. In addition to saving water, gray-water recovery systems also greatly reduce the load on the septic tank. Toilet sewage still goes to the septic system, as does kitchen water, which can contain high levels of grease and other organic materials.

In any septic system, solids and grease are trapped in the tank, where they begin to decompose through bacterial action. Gases escape through vents in the house plumbing system, and the remaining liquid flows through a system of pipes to a seepage field beneath the yard. Bacteria in the soil complete the process of decomposing the waste, leaving it harmless to the environment.

Care and Maintenance: To keep a septic system in good order, follow the procedures in the checklist at right. Most of the time, a septic system operates efficiently with little involvement from the homeowner. However, the tank must be inspected and emptied at least as often as required by the local health department—usually every 2 to 5 years.

Foul odors and slow-draining fixtures throughout the house can signal a system that has become overloaded. First check to make sure that indoor plumbing fixtures are not at fault. A leaking faucet *(pages 76-81)* or running toilet *(pages 51 and 83)* may be sending hundreds of gallons of water into the system unnecessarily.

If the fixtures are not to blame, a service firm can often correct the matter by pumping out the septic tank, cleaning the grease trap, or clearing a blocked outlet pipe. Also call in the professionals if part of the lawn remains soggy even in dry weather. Such wetness may indicate more serious problems, including compacted soil, broken pipes, or a seepage field that is too small.

Reusing Gray Water: The most common type of gray-water system provides no water treatment except to filter out lint and similar pipe-clogging debris. Because the recycled water is not treated, most communities restrict its use to subsurface irrigation, as in the system explained on pages 124 and 125. A few jurisdictions also allow the water to be piped back into the house for toilet flushing.

Upkeep for a Gray-Water System: Like a septic tank, a gray-water setup should be inspected regularly—yearly is recommended. Some systems also require the homeowner to clean or change the filter periodically. Failure to keep it clean can result in a blockage there, at the pump, or in the irrigation system. For other tips on use and maintenance of such a system, consult the checklist on page 124.

Maintenance Tips for a Septic System

✔ Repair leaking faucets and other valves as soon as possible to reduce the volume of wastewater.

✔ Avoid periods of heavy demand by staggering laundry, baths, and showers.

✔ Minimize use of chemicals such as bleach, toilet bowl cleaners, and drain openers, which upset the natural bacterial action of the system.

✔ Do not pour coffee grounds, cooking oils, fat, or grease down a drain, even if the system has a grease trap; such traps are for ordinary levels of grease and should not be overburdened.

✔ Never drive or park a car atop the seepage field; the weight may damage pipes or compact the soil, hindering absorption.

✔ When a septic system is installed, a plan is usually filed with the local code enforcement office. Obtain a copy of your home's septic system design to have on hand if major repairs are needed.

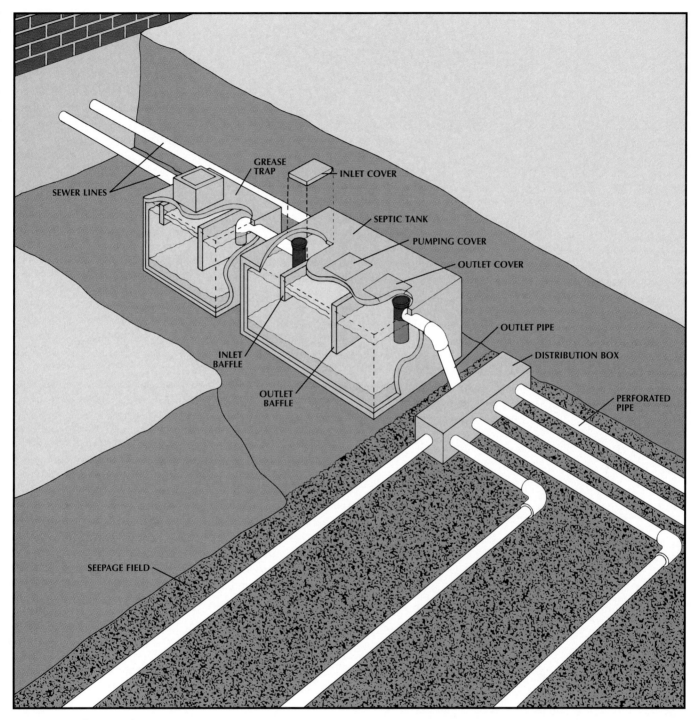

SEWER LINES

GREASE TRAP

INLET COVER

SEPTIC TANK

PUMPING COVER

OUTLET COVER

INLET BAFFLE

OUTLET BAFFLE

OUTLET PIPE

DISTRIBUTION BOX

PERFORATED PIPE

SEEPAGE FIELD

Anatomy of a septic system.

Household sewage comes from the main drain into a sewer line that leads to the septic tank; in the system shown here, a separate line carries kitchen waste through a grease trap to reduce the amount of grease entering the septic tank. Inside the tank, an inlet baffle slows the flow of waste, allowing solids to settle to the bottom, where natural bacterial action breaks them down. All septic tanks also have an inlet cover for inspection and cleaning; optional pumping and outlet covers allow additional access. An outlet baffle keeps any floating scum inside the tank as partially treated sewage passes through to the seepage field, in some cases by way of a distribution box that apportions the flow evenly. The seepage field consists of perforated pipes (or, in some systems, clay pipes with open joints) set in gravel and slanted away from the house below the frostline.

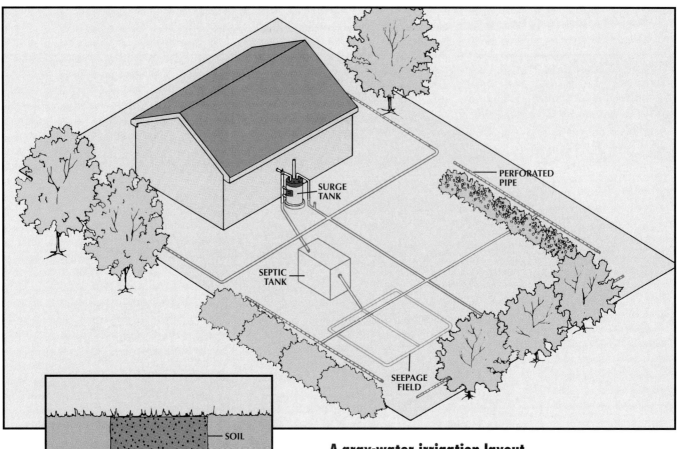

SOIL

BUILDING PAPER

PERFORATED PIPE

GRAVEL

SURGE TANK

PERFORATED PIPE

SEPTIC TANK

SEEPAGE FIELD

A gray-water irrigation layout.

Since homes produce a limited amount of wastewater, a watering system based on gray water normally serves trees, shrubs, flowers, or patches of ground cover rather than an entire lawn. In the typical arrangement above, one sewer line carries toilet and kitchen waste to the septic system while a second conveys household gray water through a filter to a surge tank. From the tank, the gray water is piped to several irrigation zones, where it is released through perforated pipes. For trees or shrubs, pipes are located at the edge of the foliage, where rain drips from leaves. Each pipe is buried at least 15 inches belowground, with 3 inches of gravel below and 2 inches above *(inset);* a layer of building paper keeps soil from settling into the gravel.

Managing a Gray-Water System

✔ Wash clothes with detergents specifically formulated to be used in a gray-water recycling system.

✔ To avoid overwatering, turn the gray-water irrigation system off when it rains.

✔ Bypass the system if wastewater is temporarily contaminated—for instance, when you are washing diapers or using bleach or caustic cleaners.

✔ Do not place plants that prefer acidic soil—azaleas, rhododendrons, hydrangeas, and begonias, for example—near a gray-water irrigation pipe. The water is high in alkalines contained in detergents.

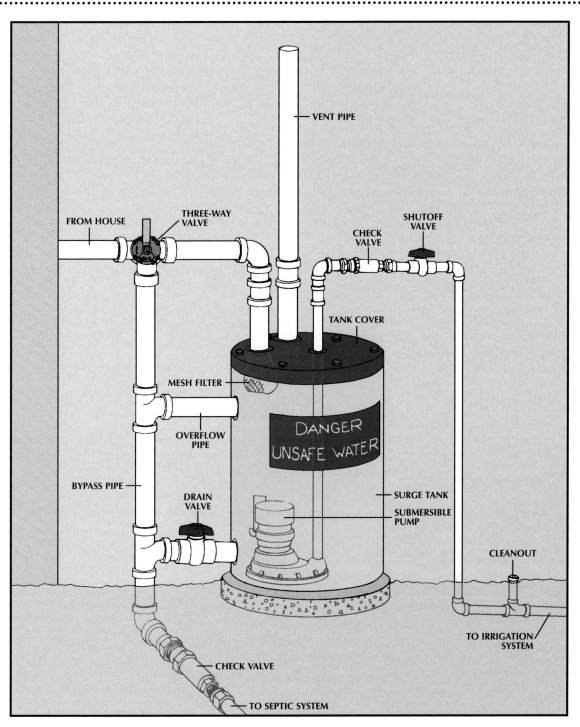

Labels in illustration:
VENT PIPE
FROM HOUSE
THREE-WAY VALVE
CHECK VALVE
SHUTOFF VALVE
TANK COVER
MESH FILTER
OVERFLOW PIPE
BYPASS PIPE
DRAIN VALVE
DANGER UNSAFE WATER
SURGE TANK
SUBMERSIBLE PUMP
CLEANOUT
TO IRRIGATION SYSTEM
CHECK VALVE
TO SEPTIC SYSTEM

The heart of the system.

In a typical gray-water recycling setup, the gray water flows out of the house through a sewer line. A three-way valve positioned on the line can redirect water to the septic system as needed—for example, during winter months or periods of heavy rains. Normally, however, water flows through the valve to a line that ends inside the surge tank, passing through a mesh filter at the end of the pipe. A removable tank cover allows access for filter replacement. Near the top of the tank, an overflow line sends excess gray water to the septic system when the tank becomes too full; at the base of the tank, another valve permits draining in case repairs are needed. When water in the tank rises, a float switch turns on a submersible pump that forces the water past a shutoff valve and on to the irrigation system. Check valves prevent water from flowing backward. As with any drain system, a cleanout plug provides access to the pipe in case of blockage *(page 43)*. By law, the tank and its associated pipes must have a warning like that shown so gray water is never confused with drinking water.

Time-Life Books is a division of Time Life Inc.

PRESIDENT and CEO: John M. Fahey Jr.
EDITOR-IN-CHIEF: John L. Papanek

TIME-LIFE BOOKS

MANAGING EDITOR: Roberta Conlan

Director of Design: Michael Hentges
Director of Editorial Operations:
  Ellen Robling
Director of Photography and Research:
  John Conrad Weiser
Senior Editors: Russell B. Adams Jr.,
  Dale M. Brown, Janet Cave, Lee Hassig,
  Robert Somerville, Henry Woodhead
Special Projects Editor: Rita Thievon
  Mullin
Director of Technology: Eileen Bradley
Library: Louise D. Forstall

PRESIDENT: John D. Hall

Vice President, Director of Marketing:
  Nancy K. Jones
*Vice President, Director of New Product
  Development:* Neil Kagan
Vice President, Book Production: Marjann
  Caldwell
Production Manager: Marlene Zack
Quality Assurance Manager: James King

HOME REPAIR AND IMPROVEMENT

SERIES EDITOR: Lee Hassig
Administrative Editor: Barbara Levitt

Editorial Staff for *Plumbing*
Senior Art Director: Cynthia Richardson
Art Director: Mary Gasperetti
Picture Editor: Catherine Chase Tyson
Text Editor: Esther Ferington
Associate Editors/Research-Writing:
  Annette Scarpitta, Karen Sweet
Technical Art Assistant: Angela Johnson
Senior Copyeditor: Juli Duncan
Copyeditor: Judith Klein
Picture Coordinator: Paige Henke
Editorial Assistant: Amy S. Crutchfield

Special Contributors: John Drummond
  (illustration); William Graves, Craig
  Hower, Marvin Shultz, Eileen Wentland
  (digital illustration); James Bowie,
  George Constable, Zachary Dorsey,
  J. T. Holland, Dean Nadalin, Peter
  Pocock, Glen Ruh, Eric Weissman (text);
  Mel Ingber (index).

Correspondents: Christine Hinze (London),
  Christina Lieberman (New York), Maria
  Vincenza Aloisi (Paris).

PICTURE CREDITS

Cover: Photograph, Renée Comet. Art,
  Carol Hilliard/Totally Incorporated.

Illustrators: Jack Arthur, Frederic F. Bigio
  from B-C Graphics, Adolph E. Brotman,
  Roger Essley, Nicholas Fasciano, Charles
  Forsythe, Donald Gates, William J. Hen-
  nessy Jr. from A & W Graphics, Fred
  Holz, John Jones, Dick Lee, Gerard
  Mariscalchi, John Martinez, Peter
  McGinn, Joan McGurren, Robert Paquet,
  Jacques Proulx, Michael Secrist, Ray
  Skibinski, Vantage Art, Inc., Whitman
  Studio, Inc.

Photographers: **End papers**: Renée Comet.
  7, 8, 9, 12, 13, 18, 23, 35: Renée
  Comet. **36**: Ken Kay. **42**: Renée Comet.
  53: LaMotte Company. **59, 61, 65**:
  Renée Comet. **70**: Moen Incorporated.
  86, 105, 106, 111, 112: Renée Comet.
  118: Rain Bird®. **121**: Renée Comet.

ACKNOWLEDGMENTS

The editors are particularly indebted to
Theresa L. Dagenhart, Long's Corporation,
Fairfax, Va., for her help in the preparation
of this volume. The editors also wish to
thank the following individuals and institu-
tions: Ted Adams, Fluid Systems, Santa
Barbara, Calif.; Allan Biggers, Charlotte
Pipe and Foundry, Charlotte, N.C.; John
Brown, Kohler Company, Kohler, Wis.;
Tom Buckley, Washington Suburban Sani-
tary Commission, Laurel, Md.; Carla Conte,
Rain Bird®, San Diego; Cheryl L. Douglas,
Long's Corporation, Fairfax, Va.; Debbie
Hartnett, Josam Company, Michigan City,
Ind.; Steve Hassett, Falls Church, Va.; Sue
Jones, Genova Products, Davison, Mich.;
Andrew G. Kireta Jr., Copper Development
Association, Crown Point, Ind.; LaMotte
Company, Chestertown, Md.; Max Limpert,
Star Water Systems, Kendallville, Ind.; Rob-
in Martin, Town & Country Baths, Washing-
ton, D.C.; Randy Millar, Highland Homes,
Inc., Trafford, Pa.; Wally Nabors, Star Water
Systems, Kendallville, Ind.; Allen E. Pfen-
ninger, Moen, Inc., North Olmsted, Ohio;
Don Pierson, Long's Corporation, Fairfax,
Va.; Nancy Rapp, Bristolpipe, Bristol, Ind.;
Denny Speas, NIBCO, Inc., Elkhart, Ind.;
Joe Teets, Fairfax, Va.; Richard Tripp, Long's
Corporation, Fairfax, Va.; Water Ace Pump
Company, Ashland, Ohio.

First printing. Printed in U.S.A.
Published simultaneously in Canada.
School and library distribution by Time-Life
Education, P.O. Box 85026, Richmond,
Virginia 23285-5026.

TIME-LIFE is a trademark of Time Warner
Inc. U.S.A.

**Library of Congress
Cataloging-in-Publication Data**
Plumbing / by the editors of Time-Life
  Books.
  p. cm. — (Home repair and improve-
  ment)
Includes index.
ISBN 0-7835-3866-9
1. Plumbing—Amateurs' manuals.
I. Time-Life Books. II. Series.
TH6124.P6 1995
696'.1—dc20   94-48025